Internet of Things Security and Privacy

The Internet of Things (IoT) concept has emerged partly due to information and communication technology developments and societal needs, expanding the ability to connect numerous objects. The wide range of facilities enabled by IoT has generated a vast amount of data, making cybersecurity an imperative requirement for personal safety and for ensuring the sustainability of the IoT ecosystem. This book covers security and privacy research in the IoT domain, compiling technical and management approaches, addressing real-world problems, and providing practical advice to the industry. This book also includes a collection of research works covering key emerging trends in IoT security and privacy that span the entire IoT architecture layers, focusing on different critical IoT applications such as advanced metering infrastructure and smart grids, smart locks, and cyber-physical systems.

The provided state-of-the-art body of knowledge is essential for researchers, practitioners, postgraduate students, and developers interested in the security and privacy of the IoT paradigm, IoT-based systems, and any related research discipline. This book is a valuable companion and comprehensive reference for postgraduate and senior undergraduate students taking an advanced IoT security and privacy course.

Internet of Things Security and Privacy

Practical and Management Perspectives

Edited by

Ali Ismail Awad, Atif Ahmad,
Kim-Kwang Raymond Choo, and Saqib Hakak

CRC Press is an imprint of the
Taylor & Francis Group, an **informa** business

First edition published 2024
by CRC Press
2385 Executive Center Drive, Suite 320, Boca Raton, FL 33431

and by CRC Press
4 Park Square, Milton Park, Abingdon, Oxon, OX14 4RN

CRC Press is an imprint of Taylor & Francis Group, LLC

© 2024 selection and editorial matter, [Ali Ismail Awad, Atif Ahmad, Kim-Kwang Raymond Choo, Saqib Hakak]; individual chapters, the contributors

Reasonable efforts have been made to publish reliable data and information, but the author and publisher cannot assume responsibility for the validity of all materials or the consequences of their use. The authors and publishers have attempted to trace the copyright holders of all material reproduced in this publication and apologize to copyright holders if permission to publish in this form has not been obtained. If any copyright material has not been acknowledged please write and let us know so we may rectify in any future reprint.

Except as permitted under U.S. Copyright Law, no part of this book may be reprinted, reproduced, transmitted, or utilized in any form by any electronic, mechanical, or other means, now known or hereafter invented, including photocopying, microfilming, and recording, or in any information storage or retrieval system, without written permission from the publishers.

For permission to photocopy or use material electronically from this work, access www.copyright.com or contact the Copyright Clearance Center, Inc. (CCC), 222 Rosewood Drive, Danvers, MA 01923, 978-750-8400. For works that are not available on CCC please contact mpkbookspermissions@tandf.co.uk

Trademark notice: Product or corporate names may be trademarks or registered trademarks, and are used only for identification and explanation without intent to infringe.

ISBN: 9781032057712 (hbk)
ISBN: 9781032058306 (pbk)
ISBN: 9781003199410 (ebk)

DOI: 10.1201/9781003199410

Typeset in Times
by codeMantra

Contents

Preface xi

Editors xv

Contributors xvii

Chapter 1 Cybersecurity Risk Assessment in Advanced Metering Infrastructure 1

Mostafa Shokry Abd El Salam, Ali Ismail Awad, Mahmoud Khaled Abd-Ellah, and Ashraf A. M. Khalaf

1.1 Introduction 1
1.2 Preliminaries 3
1.2.1 Advanced Metering Infrastructure 3
1.2.2 AMI Components 4
1.2.3 AMI Tiers 5
1.2.4 Information Security Risk Assessment 6
1.3 Implementation of the AMI System's Risk Assessment 9
1.3.1 Risk Identification Phase for the AMI System 9
1.3.2 AMI Vulnerabilities 11
1.3.3 Risk Profiling Phase for the AMI System 14
1.3.4 Risk Treatment Phase for the AMI System 16
1.4 Discussion and Recommendations 17
1.4.1 Recommendations 18
1.5 Conclusion 19
Acknowledgment 19
References 20

Chapter 2 A Generative Neural Network for Improving Metamorphic Malware Detection in IoT Mobile Devices 24

Leigh Turnbull, Zhiyuan Tan, and Kehinde O. Babaagba

2.1 Introduction 24
2.2 Background 26
2.2.1 Machine Learning 26
2.2.2 Deep Learning Malware Detection 28
2.2.3 Adversarial Machine Learning 30
2.2.4 Generative Adversarial Networks 31

2.2.5 Related Work 33
2.3 Methodology 35
2.3.1 Dataset 35
2.3.2 Dynamic Analysis 35
2.3.3 Data Preparation 36
2.3.4 Image Generation 37
2.3.5 Adversarial Samples 37
2.3.6 Convolutional Neural Network (CNN) 38
2.4 Experimental Design 38
2.4.1 Experimental Setup 39
2.4.2 Behavior Feature Extraction 40
2.4.3 Words to Images 41
2.4.4 Synthetic Images 41
2.4.5 Image Classification 43
2.5 Results and Discussion 45
2.5.1 Assessing the Evasive Effectiveness of the Generated Samples Using a CNN Classifier 45
2.5.2 Assessing the Effectiveness of the CNN Classifier with a Novel Dataset Including a Newly Generated Batch of Malicious Samples for Each Family Produced by the DCGAN 46
2.5.3 Evaluation 47
2.6 Conclusion 48
Notes 48
References 49

Chapter 3 A Physical-Layer Approach for IoT Information Security During Interference Attacks 54

Abdallah Farraj and Eman Hammad

3.1 Introduction 54
3.2 Chapter Contributions 56
3.3 Related Work 57
3.4 IoT Information Security 57
3.4.1 Background 58
3.4.2 System Model 58
3.5 Zero-Determinant Strategies 59
3.6 Game-Theoretic Transmission Strategy 61
3.6.1 Transmission Probability 61
3.6.2 Transmission Strategy 63
3.7 Extension to Multiple IoT Users 65
3.7.1 Zero-Determinant Strategies 65
3.7.2 Generalized Transmission Strategy 66
3.8 Numerical Results 66

3.8.1 Model Dynamics 66
3.8.2 Simulated Use Cases 68
3.9 Discussions 68
3.9.1 About the Game-Theoretic Approach 68
3.9.2 Conclusions 73
References 73

Chapter 4 Policy-Driven Security Architecture for Internet of Things (IoT) Infrastructure 76

Kallol Krishna Karmakar, Vijay Varadharajan, and Uday Tupakula

4.1 Introduction 76
4.2 Related Work 79
4.2.1 Policies and SDN 79
4.2.2 Automatic Device Provisioning 79
4.2.3 Secure Device Provisioning 80
4.2.4 Machine Learning-based Classification of Devices 81
4.2.5 IoT Security and Attacks 81
4.3 Fundamentals of Policy-Based Network and Security Management 82
4.3.1 Policy 82
4.3.2 Policy-Based Network and Security Management 82
4.3.3 Policy-Based Management Architecture 83
4.3.4 Benefits of a Policy-Based Management Architecture 84
4.4 IoT Network Scenario 85
4.4.1 Types of Devices and Device Ontology 86
4.5 Policy-Driven Security Architecture 89
4.5.1 Device Provisioning? 89
4.5.2 Secure Smart Device Provisioning and Monitoring Service (SDPM) 94
4.5.3 Security Provisioning Protocol 95
4.5.4 Digital Twin 96
4.5.5 Policy-Based Security Application 98
4.6 Prototype Implementation 100
4.6.1 Network Setup 100
4.6.2 Security Analysis 101
4.6.3 Performance Evaluation 109
4.7 Discussion and Open Issues 114
4.8 Conclusion 116
References 117

Chapter 5 A Privacy-Sensitive, Situation-Aware Description Model for IoT 121

Zakaria Maamar, Amel Benna, Noura Faci, Fadwa Yahya, Nacereddine Sitouah, and Wassim Benadjel

5.1 Introduction 121
5.2 Background 123
5.2.1 Privacy in IoT in-Brief 123
5.2.2 Definitions 124
5.2.3 When MDA Meets IoT 124
5.2.4 WoT TD in-Brief 126
5.2.5 Case Study 127
5.3 Privacy-Sensitive and Situation-Aware Thing Description 128
5.3.1 Overview 128
5.3.2 Step 1: SituationPrivacy Metamodel Definition 129
5.3.3 Step 2: SituationPrivacyWoTTD Metamodel Definition 131
5.3.4 Step 3: SituationPrivacyWoTTD Model Generation 132
5.4 Implementation 132
5.4.1 Model Transformation 133
5.4.2 Simulation 137
5.4.3 Evaluation 138
5.5 Conclusion 140
Appendix 1 141
Notes 141
References 142

Chapter 6 Protect the Gate: A Literature Review of the Security and Privacy Concerns and Mitigation Strategies Related to IoT Smart Locks 146

Hussein Hazazi and Mohamed Shehab

6.1 Introduction 146
6.1.1 Background 147
6.1.2 Architecture 147
6.1.3 Capabilities 149
6.1.4 Access Control 150
6.1.5 Authentication and Authorization 150
6.2 The Privacy and Security of Smart Locks 151
6.2.1 Smart Locks Privacy and Security from the Perspective of Researchers 151
6.2.2 Smart Homes Privacy and Security from the Perspective of the End User 155

6.3 Research Gaps ... 160
6.4 Conclusion ... 161
References ... 162

Chapter 7 A Game-Theoretic Approach to Information Availability in IoT Networks ... 165

Abdallah Farraj and Eman Hammad

7.1 Introduction ... 165
7.2 Related Work ... 167
7.3 System Model ... 168
7.3.1 Spectrum-Sharing Cognitive Systems ... 168
7.3.2 Problem Statement ... 168
7.3.3 Primary Outage Probability ... 170
7.4 Zero-Determinant Strategies ... 171
7.5 Game-Theoretic Strategy for IoT Transmission ... 173
7.5.1 Uncoordinated Transmission Strategy ... 173
7.5.2 Special Cases ... 176
7.5.3 Performance Analysis ... 177
7.6 Extension to Multiple Users ... 179
7.7 Numerical Results ... 180
7.8 Discussions and Conclusions ... 182
References ... 185

Chapter 8 Review on Variants of Restricted Boltzmann Machines and Autoencoders for Cyber-Physical Systems ... 188

Qazi Emad Ul Haq, Muhammad Imran, Kashif Saleem, Tanveer Zia, and Jalal Al Muhtadi

8.1 Introduction to RBMs and Autoencoding ... 188
8.2 Background ... 189
8.2.1 Targeted Problems Using RBM's and Autoencoders ... 189
8.2.2 Techniques Used for Cyber-Physical Systems Using RBMs and Autoencoders ... 191
8.2.3 Detecting Network Intrusions to Ensure the Security of CPS in IoT Devices ... 194
8.3 Malware Attack Detection ... 195
8.4 Fraud and Anomaly Detection ... 196
8.5 Breakthroughs in CPS and Their Findings ... 199
8.5.1 Aim of a CPS-Based System ... 200
8.5.2 Breakthroughs in CPS-Based Systems ... 200
8.6 Ensuring CPS is Critical in the Modern World ... 201
8.7 Evolution of CPS and Its Associated Impacts ... 202
8.8 Conclusion ... 204

Acknowledgment 205
References 205

Chapter 9 Privacy-Preserving Analytics of IoT Data Using Generative Models 208
Magd Shareah, Rami Malkawi, Ahmed Aleroud, and Zain Halloush

9.1 Introduction 208
9.2 IoT Architecture and Applications 209
9.3 Limitations and Challenges 210
9.4 IoT Privacy: Definitions and Types 212
9.5 GAN Framework 212
9.6 Research Objectives 214
9.6.1 Limitation of the Scope 214
9.7 Literature Review 215
9.7.1 Data Anonymizing 215
9.7.2 Authentication and Authorization 217
9.7.3 Edge Computing and Plug-in Architecture 219
9.7.4 Using Generative Adversarial Network (GAN) in Privacy Data Analytics 220
9.8 Overall Research Design 221
9.9 Methodology 222
9.9.1 Data Preparation 223
9.10 Data Analysis and Interpretation 224
9.10.1 Privacy Measures 225
9.10.2 Accuracy Measures 225
9.10.3 Incorrect Classification 225
9.10.4 F-Measure 225
9.10.5 Privacy 225
9.10.6 Privacy Results Using Different Number of Epochs 226
9.11 Conclusion and Future Work 228
References 228

Index 233

Preface

The Internet of Things (IoT) is a nascent paradigm that has emerged due to significant advancements in information and communication technology (ICT). Its primary goal is to enhance the functionality of the original version of the Internet by facilitating connections between an array of objects. The IoT model has since evolved to encompass various applications, including Industry 4.0 and manufacturing systems, cyber-physical systems, eHealth, smart cities and smart homes, robotics and drones, transportation, and critical infrastructures. Despite the many advantages of IoT and other sensing technologies, they generate a massive volume of data across various domains, necessitating security and privacy measures to safeguard personal safety and ensure the sustainability of the IoT paradigm. Moreover, the nature and importance of IoT systems themselves make them attractive targets for attack. Therefore, achieving the highest levels of security and privacy is crucial for reaping the full benefits of IoT systems. However, owing to the broad range of IoT applications and environments, several security and privacy concerns still need to be addressed.

This book addresses the gaps in IoT security and privacy by providing cutting-edge research findings in the IoT security and privacy domains. The uniqueness of this book volume emerges from combining both practical and management viewpoints in one place. This book outlines the latest emerging trends in IoT security and privacy from practical and management perspectives, focusing on the entire IoT architecture, including the perception, network, application layers, and critical IoT applications. The provided up-to-date body of knowledge is essential for researchers, practitioners, and postgraduate students involved in IoT security, development, and deployment. This book comprises nine chapters written by experts in the field, covering both the security and privacy aspects of IoT. The material is presented in a way that allows each chapter to be read independently while contributing to a collective understanding of the topic. The nine chapters of this book is organized as follows:

The book volume begins with a chapter titled "Cybersecurity Risk Assessment in Advanced Metering Infrastructure," authored by *Shokry et al.*, which contributes to the risk assessment of advanced metering infrastructure (AMI) as a category of IoT and a component of smart grids. The AMI system is a type of the Internet of Things (IoT) technology that is increasingly used today. Its purpose is to collect data on the electricity consumption of customers and transmit this information to electricity service providers (ESP) for storage, processing, and analysis. Integrating information and communication technology (ICT) with the conventional electric power grid has posed new security challenges for the AMI system. To enhance the security of this system, one of the precautions that can be taken is using a risk assessment process. The contribution of this chapter is an evaluation of the current cybersecurity risks to the AMI system. By assessing these risks, it is possible to identify critical assets that require protection, potential vulnerabilities, the likelihood of threats, and their potential impacts. The authors establish a risk matrix to evaluate the potential hazards and select the best risk management approach. This study identifies nine risks for each AMI layer and component. The degree of risk, likelihood, and impact of each risk are determined. Finally, the authors propose a mitigation approach to reduce the risk level to a manageable degree.

Nowadays, artificial intelligence and machine learning play an influential role in cybersecurity, especially in attack detection, prediction, and prevention. Chapter 2, titled "A Generative Neural Network for Improving Metamorphic Malware Detection in IoT Mobile Devices," authored by *Turnbull, Tan, and Babaagba*, addresses the malware detection problem by applying a generative neural network (GNN) approach. The chapter discusses the increasing occurrence of malicious attacks on computer systems and networks, specifically the emergence of new malware families targeting information assets. One such group is metamorphic malware, which employs multiple obfuscation techniques to alter its code structure between generations, making detection and analysis more challenging. The research presented in this chapter focuses on improving the detection of metamorphic malware in the Android operating system (OS) by augmenting training data with new samples generated through deep convolutional generative adversarial networks (DCGAN) and features from existing metamorphic malware samples. Experimental results demonstrate improved detection of novel metamorphic malware through this method.

Chapter 3, titled "A Physical-Layer Approach for IoT Information Security During Interference Attacks," written by *Farraj and Hammad*, considers a heterogeneous communication environment with IoT devices transmitting over a wireless channel in the presence of adversarial IoT devices inducing jamming interference attacks. The chapter discusses a communication environment involving IoT devices transmitting wirelessly, with some IoT devices intentionally causing jamming interference attacks. This chapter proposes using game theory and an iterated game formulation to develop a physical-layer security approach to ensure that IoT devices can still access information despite active interference. To reduce the scheduling overhead, a game-theoretic transmission strategy for uncoordinated IoT channel access is considered, which achieves the desired security metric while conserving IoT device resources. The interactions between a representative IoT device and adversary devices are investigated to determine the effect of intentional interference on signal quality and quantify the IoT information availability. The need for more efficient and scalable scheduling approaches for large-scale IoT deployments is also considered. Simulation results illustrate how a selected IoT device can achieve its desired security performance over time using the proposed transmission strategies.

Chapter 4, titled "Policy-Driven Security Architecture for Internet of Things (IoT) Infrastructure," prepared by *Karmakar, Varadharajan, and Tupakula*, introduces a policy-driven security architecture that consists of two major services: Secure smart device provisioning and monitoring service architecture (SDPM) and policy-based security application (PbSA). The security architecture suits programmable smart network infrastructures such as IoT-enabled smart homes or offices, industrial IoT infrastructures, and healthcare infrastructure. The SDPM service in the architecture allows for the provisioning of devices to control the activities of malicious devices using a dynamic policy-based approach. It provides pre- and post-condition-based policies to provision IoT devices securely and control their runtime operations. The digital twin concept is used to represent the security status of the devices, which is used for dynamic security status monitoring at runtime, automating the update and patch management on-demand. Additionally, PbSA helps enforce fine-grained policies to secure the flows in the IoT network infrastructure, creating a secure network infrastructure for IoT devices.

In addition to security, privacy is another concern in IoT deployments. Chapter 5, "A Privacy-Sensitive, Situation-Aware Description Model for IoT," authored by *Maamar et al.*, addresses the security privacy challenge in the IoT paradigm. In this chapter, the impact of IoT on people's private lives is discussed, and the expected benefits of IoT have become a concern for people. To address this concern, the chapter proposes a model-driven architecture (MDA)-based approach to incorporate privacy concerns into thing specifications. This approach makes things aware of privacy concerns and enables them to act accordingly, improving their assessment of what needs to be done before violating privacy policies. The chapter uses WoT thing description (WoT TD) as an example of a W3C thing specification to demonstrate the approach. The approach defines the SituationPrivacy metamodel, merges it with the WoT TD metamodel to create the SituationPrivacyWoTTD metamodel, and generates the SituationPrivacyWoTTD model using an in-house case tool. The chapter also provides a case study about a care center for elderly people to illustrate and demonstrate privacy concerns and the MDA approach used to address them.

In recent years, smart home devices have become increasingly popular among homeowners. The smart lock is one of the smart home devices whose market size has increased significantly. Chapter 6, titled "Protect the Gate: A Literature Review of the Security and Privacy Concerns and Mitigation Strategies Related to IoT Smart Locks," written by *Hazazi and Shehab*, discusses smart locks, which are smart devices that use IoT-enabled sensors to allow keyless entry to residential and commercial facilities. Unlike traditional locks, smart locks can be remotely operated using a smartphone and offer additional features and functionalities. However, due to their location within the home, it is essential to examine their security and privacy thoroughly. The chapter presents a literature review investigating the security and privacy concerns related to smart locks and the strategies proposed to mitigate these concerns. The review considers both the perspective of researchers, based on their analysis of smart lock security and privacy, and the perspective of end-users, based on their day-to-day experience with smart home devices.

Chapter 7, titled "A Game-Theoretic Approach to Information Availability in IoT Networks," written by *Farraj and Hammad*, considers the information availability for IoT devices. The chapter proposes a physical-layer security approach to protect information availability for IoT devices. The proposed approach uses a game-theoretic-based distributed transmission strategy in an industrial sensor network environment where IoT devices compete to transmit their measurements over a shared channel. The approach aims to satisfy quality of service (QoS) requirements while minimizing coordination overhead. The proposed model focuses on one resource-constrained IoT device that can transmit over the channel during specific intervals. The game-theoretic strategy employs an iterated zero-determinant formulation to model interactions between IoT devices and develop an uncoordinated channel access strategy that satisfies QoS constraints and achieves information availability objectives. Simulation results show that the proposed approach can achieve target performance over time without the need for coordinated channel transmissions with other IoT devices, provided certain conditions are satisfied.

Cyber-physical systems (CPS) are one of the application domains where the IoT paradigm is deployed. Thus, CPS and IoT are linked in terms of security and privacy. Chapter 8, titled "Review on Variants of Restricted Boltzmann Machines and Autoencoders for Cyber-Physical Systems," prepared by *Emad Ul Haq et al.*, discusses two types of neural networks: Restricted Boltzmann machines (RBMs) and autoencoders, their variants, and their applications in CPS. RBMs are generative and can generate new data, while autoencoders are used for dimensionality reduction. CPS can use RBMs to identify hidden states and minimize system energy. Autoencoders are unsupervised and study compressed data encoding autonomously. The chapter also discusses breakthroughs in CPS, highlighting the limitations of previous evaluation techniques and the importance of artificial intelligence design and analysis as an integral part of CPS. Overall, the chapter provides insights into the advancements in RBMs, autoencoders, and CPS and their potential impact on various fields, including safety and artificial intelligence.

The book volume ends with Chapter 9, titled "Privacy-Preserving Analytics of IoT Data Using Generative Models," authored by *Shareah et al.*, which discusses the rapid growth of the IoT and its impact on our daily lives. The IoT allows for data collection and sharing but also raises security concerns. The growing number of connected devices implies instant surveillance of almost every action, sound, and move we encounter daily. This has resulted in a decline in the efficiency and performance of existing privacy-preserving methods. To address this, the chapter proposes a generative privacy-preserving model for IoT data. The proposed model perturbs data while preserving its features, and evaluation metrics show remarkable results in terms of accuracy and privacy.

Ultimately, this book comprehensively covers the latest advancements, contemporary challenges, and pioneering research discoveries pertaining to IoT security and privacy across various domains from practical and management perspectives. It serves as a testament to the significant progress made in this field of study, which is poised for continued growth in light of the rapid expansion of IoT technologies. While additional insights and developments are expected in this area, the contributions featured in this book offer valuable ideas and insights that can aid in various contexts and promote a holistic comprehension of IoT security and privacy. We trust that readers will find this book engaging and relevant and a valuable addition to the existing body of IoT security and privacy literature.

Ali Ismail Awad, United Arab Emirates University, Al Ain, United Arab Emirates
Atif Ahmad, The University of Melbourne, Melbourne, Australia
Kim-Kwang Raymond Choo, The University of Texas at San Antonio,
San Antonio, Texas, USA
Saqib Hakak, University of New Brunswick, New Brunswick, Canada

Editors

Ali Ismail Awad (Ph.D., SMIEEE, MACM) is currently an associate professor with the College of Information Technology (CIT), United Arab Emirates University (UAEU), Al Ain, United Arab Emirates. He has been coordinating the Master's Program in Information Security at the same institution since 2022. Dr. Awad is an associate professor with the Electrical Engineering Department, Faculty of Engineering, Al-Azhar University at Qena, Egypt. He is also a visiting researcher at the University of Plymouth, United Kingdom. Dr. Awad was an associate professor (Docent) with the Department of Computer Science, Electrical and Space Engineering, LuleåUniversity of Technology, Luleå, Sweden, where he coordinated the Master's Program in Information Security from 2017 to 2020. His research interests include cybersecurity, network security, IoT security, and image analysis with biometrics and medical imaging applications. Dr. Awad has edited or co-edited several books and authored or co-authored several journal articles and conference papers in these areas. He is an editorial board member of the *Future Generation Computer Systems Journal*, *Computers & Security Journal*, *The Internet of Things*, *Engineering Cyber-Physical Human Systems Journal*, *The Health Information Science and Systems Journal*, and *The Security, Privacy and Authentication* section of *Frontiers*. Dr. Awad is currently an IEEE senior member and an ACM Professional Member.

Atif Ahmad (Ph.D., CPP) is an associate professor at the University of Melbourne's School of Computing & Information Systems where he serves as Deputy Director of the Academic Centre of Cyber Security Excellence. Atif leads a unique team of Cybersecurity Management researchers drawn from information systems, business administration, security intelligence, and information warfare. He has authored over 100 scholarly articles in cybersecurity management including 20 journal articles ranked in Q1, the top quartile of quality. Atif has received over AUD$5M in grant funding from the Australian Research Council, the Australian Department of Foreign Affairs and Trade, the Menzies Foundation, and Australian Defence. Atif is an associate editor for the leading IT security journal, *Computers & Security*. Atif has developed and co-produced multi-award-winning films (including "best original screenplay" at the London International Film Festival) showcasing cyber practices in the industry. He has previously served as a cybersecurity consultant for WorleyParsons, Pinkerton, and SinclairKnightMerz. Atif is a certified protection professional with the American Society for Industrial Security. For more information, please visit https://www.atifahmad.me/.

Kim-Kwang Raymond Choo (Ph.D.) received the Ph.D. in Information Security in 2006 from Queensland University of Technology, Australia. He currently holds the Cloud Technology Endowed Professorship at The University of Texas at San Antonio. He is the founding co-editor-in-chief of ACM Distributed Ledger Technologies: Research & Practice, and the founding chair of IEEE Technology and Engineering Management Society Technical Committee (TC) on Blockchain and Distributed Ledger Technologies. He is the recipient of the 2022 IEEE Hyper-Intelligence TC Award for Excellence in Hyper-Intelligence Systems (Technical Achievement award), the 2022 IEEE TC on Homeland Security Research and Innovation Award, the 2022 IEEE TC on Secure and Dependable Measurement Mid-Career Award, and the 2019 IEEE TC on Scalable Computing Award for Excellence in Scalable Computing (Middle Career Researcher).

Saqib Hakak (Ph.D.) is currently a tenure-track assistant professor with the Faculty of Computer Science, Canadian Institute for Cybersecurity (CIC), University of New Brunswick (UNB). He has more than five years of industry and academic experience. He has recently made it to the list of the top 2% of world researchers list for the year 2022, compiled annually by Stanford University. His current research interests include cybersecurity, data mining, applications of artificial intelligence, and emerging technologies. He has received a number of gold/silver awards in international innovation competitions. He is currently serving as an associate editor and a technical committee member/reviewer of several reputed conferences.

Contributors

Mahmoud Khaled Abd-Ellah
Department of Artificial Intelligence
Egyptian Russian University
Cairo, Egypt

Ahmed Aleroud
School of Computer and Cyber Sciences
Augusta University
Augusta, Georgia

Ali Ismail Awad
College of Information Technology
United Arab Emirates University
Al Ain, United Arab Emirates
and
Faculty of Engineering
Al-Azhar University
Qena, Egypt

Kehinde O. Babaagba
School of Computing, Engineering and Built Environment
Edinburgh Napier University
Edinburgh, Scotland, United Kingdom

Wassim Benadjel
Department of Computer Science
University of Science and Technology Houari Boumédiene
Algiers, Algeria

Amel Benna
Department of Multimedia and Information Systems
Research Center for Scientific and Technical Information (CERIST)
Algiers, Algeria

Noura Faci
Laboratoire d'InfoRmatique en Image et Systémes d'information
University of Lyon, UCBL, CNRS, INSA Lyon, LIRIS
Lyon, France

Abdallah Farraj
Department of Electrical Engineering
Texas A&M University
College Station, Texas

Zain Halloush
School of Computer and Cyber Sciences
Augusta University
Augusta, Georgia

Eman Hammad
Department of Engineering Technology and Industrial Distribution
Texas A&M University
College Station, Texas

Qazi Emad Ul Haq
Center of Excellence in Cybercrimes and Digital Forensics (CoECDF)
Naif Arab University for Security Sciences (NAUSS)
Riyadh, Saudi Arabia

Hussein Hazazi
College of Computing and Informatics
University of North Carolina at Charlotte (UNCC)
Charlotte, North Carolina

Muhammad Imran
Institute of Innovation, Science, and Sustainability
Federation University
Brisbane, Australia

Kallol Krishna Karmakar
Advanced Cybersecurity Engineering Research Centre
University of Newcastle
Newcastle, Australia

Ashraf A. M. Khalaf
Department of Electrical Engineering
Faculty of Engineering
Minia University
Minia, Egypt

Zakaria Maamar
College of Computing and IT
University of Doha for Science and Technology
Doha, Qatar

Rami Malkawi
Department of Information Systems
Yarmouk University
Irbid, Jordan

Jalal Al Muhtadi
Center of Excellence in Information Assurance (CoEIA)
King Saud University
Riyadh, Saudi Arabia

Mostafa Shokry Abd El Salam
Department of Infrastructure and Information Security
Ministry of Electricity and Renewable Energy
Cairo, Egypt

Kashif Saleem
Center of Excellence in Information Assurance (CoEIA)
King Saud University
Riyadh, Saudi Arabia

Magd Shareah
Department of Information Systems
Yarmouk University
Irbid, Jordan

Mohamed Shehab
College of Computing and Informatics
University of North Carolina at Charlotte (UNCC)
Charlotte, North Carolina

Nacereddine Sitouah
Department of Electronics, Information, and Bioengineering
Polytechnic University of Milan
Milan, Italy

Zhiyuan Tan
School of Computing, Engineering and Built Environment
Edinburgh Napier University
Edinburgh, Scotland

Uday Tupakula
School of Information and Physical Sciences
University of Newcastle
Newcastle, Australia

Leigh Turnbull
School of Computing, Engineering and Built Environment
Edinburgh Napier University
Edinburgh, Scotland

Vijay Varadharajan
Global Innovation Chair in Cyber Security
School of Information and Physical Sciences
Advanced Cybersecurity Engineering Research Centre
University of Newcastle
Newcastle, Australia

Fadwa Yahya
College of Sciences and Humanities Al-Afla
Prince Sattam bin Abdulaziz University
Al-Kharj, Saudi Arabia
and
Multimedia, Information Systems &
Advanced Computing Laboratory
University of Sfax
Sfax, Tunisia

Tanveer Zia
Department of Forensic Sciences
Naif Arab University for Security Sciences (NAUSS)
Riyadh, Saudi Arabia

1 Cybersecurity Risk Assessment in Advanced Metering Infrastructure

Mostafa Shokry Abd El Salam
Ministry of Electricity and Renewable Energy, Egypt

Ali Ismail Awad
United Arab Emirates University, Al-Azhar University

Mahmoud Khaled Abd-Ellah
Egyptian Russian University

Ashraf A. M. Khalaf
Minia University

1.1 INTRODUCTION

The backbone of the smart grid (SG) is the advanced metering infrastructure (AMI) system [1]. The data communication technologies and the conventional electrical grid were combined to create the AMI system. The AMI system improves the traditional power grid's functionality, which is reflected in the end customer. The AMI system benefits can be summarized as follows: the bidirectional communication links, the ability to notify the customer when reaching a threshold electricity consumption value, making the maintenance process for the smart meters (SMs) easier, and reporting power outages to the electricity service provider (ESP) immediately [2].

The Internet of Things (IoT) is a revolutionary, highest-rated technology that connects numerous wired or wireless devices or things to the internet for data exchange [3]. IoT technology improves the communication between entities, people, or objects that are physically separated from one another. IoT technology is now widely used in a variety of industries, including transportation, education, utilities, and business development, as shown in Figure 1.1 [4]. Currently, there are around 26.66 billion IoT devices worldwide, including smart energy meters, wearable technology, and home automation systems [5].

DOI: 10.1201/9781003199410-1

The AMI system's primary goal is to gather, store, and analyze data on electricity usage from the end customer to the utility center [6]. Smart meters (SMs), data concentrators (DCs), meter data management systems (MDMs), and communication channels utilized for data traversal are the four primary parts of the AMI system. SM is in charge of gathering information from the intelligent end devices (IED) situated in smart houses. These data show the customers' electricity usage and the power signatures of the devices used in smart homes. These data are subsequently transmitted to the DC, which is in charge of gathering the data from various SMs. Finally, the MDMs of the utility center (UC) receive the data that the DC has collected and will store and analyze. All of these data are moved across the AMI system via wired or wireless communication links [7].

Based on the preceding descriptions of the IoT and the AMI systems, as demonstrated in Figure 1.1, it is demonstrated that the AMI system is one of the IoT system examples that are currently in use. The SMs, DCs, and MDMs end devices that make up the AMI system are linked through bidirectional communication cables, so that they can exchange data with each other.

Lack of user knowledge, lack of IoT device upgrades (SM and DC in the case of the AMI system), and the wireless communication channel all exacerbated the AMI system's vulnerability to cyber adversities and dangers [8]. The security pillars, which include confidentiality, integrity, and availability (CIA) triad, target the data collected and transferred in the AMI system. Unauthorized access to any component

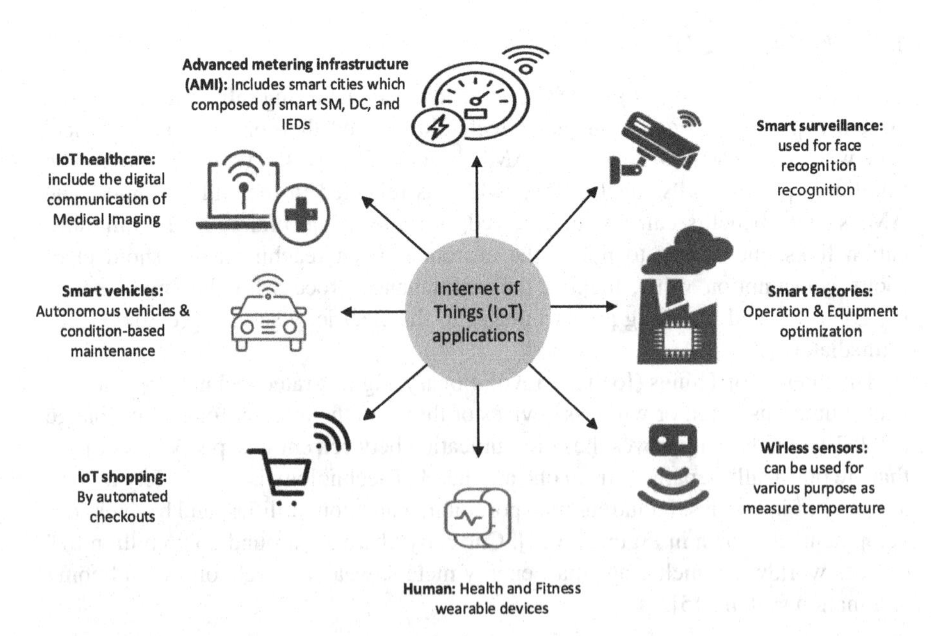

Figure 1.1 Samples of IoT applications include AMI systems, smart factories, healthcare, and smart surveillance.

of the AMI system can significantly affect these pillars, which are fundamental to the AMI security [9].

Any cyberattacks that take advantage of one of the AMI system's weaknesses could cause a partial or widespread blackout, depending on the significance of the data passing through it. For instance, on December 23, 2015, enemies gained remote access to the control center of three Ukrainian electrical distribution companies and took over the systems for monitoring, controlling, and gathering data. More than 200,000 customers lost power as a result of malicious actors opening breakers at about 30 distribution substations in Kyiv and the western Ivano-Frankivsk area [10].

Based on the ISO/IEC 27001 Standard for Information Technology — Security Techniques — Information Security Management Systems — Requirements, the procedure of risk assessment is implemented to protect the critical information system from cyberattacks [11]. The risk assessment strategy may be applied to determine the primary critical assets and the current vulnerabilities associated with these assets, and the threat agent associated with the AMI system must be identified by evaluating the possible risks of the AMI system. The risk assessment process is also necessary to demonstrate the likelihood, potential consequences, and necessary mitigation measures in order to reduce the risk level [12].

The definition of the AMI system is that AMI system is one of the existing critical information systems and its operation depends on gathering electricity consumption data. Thus, the risk assessment process is crucial for the AMI system's security. The major goal of this chapter is to carry out a security risk assessment of the AMI system, which includes all steps of the risk assessment process. This objective should enhance the existing security policies for the IoT-based AMI system.

The rest of this chapter is organized as follows: Section 1.2 covers the two pillars of our work, which are the AMI system and information security risk assessment. This section illustrates the essential elements, the objective, and the AMI system's tiers. Furthermore, it demonstrates the information security risk assessment objective, including the required steps for performing the risk assessment process and the different risk analysis approaches. Section 1.3 shows the results of assessing the potential cybersecurity risks of the AMI system. This section determines the existing critical assets, vulnerabilities, the probability in which the attack will occur, the impact of these attacks on the AMI system, and the risk levels associated with samples of attack scenarios. Section 1.4 provides an explanation for the findings in the preceding section. Furthermore, samples of mitigation approaches are recommended to mitigate the obtained security risk scenarios. Finally, Section 1.5 provides a conclusion of our work.

1.2 PRELIMINARIES

1.2.1 ADVANCED METERING INFRASTRUCTURE

AMI is one of today's most important critical information infrastructure systems, and it can be seen as the first step in digitizing the conventional electricity grid [13]. Its primary goal is to gather customer power consumption data from both SMs and DCs, which will then be transmitted over bidirectional communication lines to MDMs for storage and analysis [14].

The AMI system is an improvement of the automated meter reading (AMR) system. The AMR system is used to transfer the traditional electricity meter's energy usage to the electricity service provider (ESP) over a unidirectional communication channel. The AMI system introduced a bidirectional communication channel between the SM and the ESP that improves the functionality of the AMR system. The end user's electricity consumption data are collected through these two-way communication channels, and remote commands are sent to the SMs to carry out specified tasks, including upgrading the firmware, establishing connections, and exchanging security keys [15].

1.2.2 AMI COMPONENTS

The AMI system comprises four main components, namely, SMs, DCs, MDMs, and the communication channels between them, as shown in Figure 1.2. The major purpose of SM, the first AMI system component, is to collect data on customer-specific electricity use [16]. SM consists of two basic components, the first of which is concerned with monitoring the user's energy usage. The second component is in charge of carrying out specialized tasks, including remotely updating the firmware, connecting or detaching from a distance, identifying, and averting electricity thievery [17].

The DC, a device serving as a gateway between the SMs and the MDMs in the UC, is the following part of the AMI system [18]. DC does two basic tasks. The first is gathering data from various SMs and sending them to the MDMs. The second one involves accepting commands from MDMs and delivering them to the SMs for a specified activity [19].

The MDMs, a software and hardware combination located in the ESP's data center, are the next part of the AMI system. The MDMs' tasks can be summed up as collecting information from SMs, directing SMs to perform certain tasks via the DC, and monitoring various AMI components, including the transmission network, the centers, and the power that is produced [20].

In the AMI system, communication networks are crucial, which is the last component of the AMI system. They are in charge of tying together all of an AMI's major parts, including the MDMs, DCs, and SMs [14]. Two-way communication pathways are employed to deliver instructions to the SMs from the UC as well as commands to both DCs and MDMs from SMs [21]. Although the communication methods utilized in an AMI differ greatly, they may typically be divided into wired and wireless types. Because of the quantity and importance of the information sent through the route, using the appropriate communication mechanism is crucial, and it is essential to safeguard the AMI [22].

The hardware layer, data layer, and communication layer are the three sub-layers of the AMI system [23] as shown in Figure 1.2. The AMI hardware layer includes SM, DC, and MDM, which is located in the service provider data center [23].

The data gathered from the IED to SM are included in the data layer of the AMI system. These data contain the unique identifier for each device and the customer's

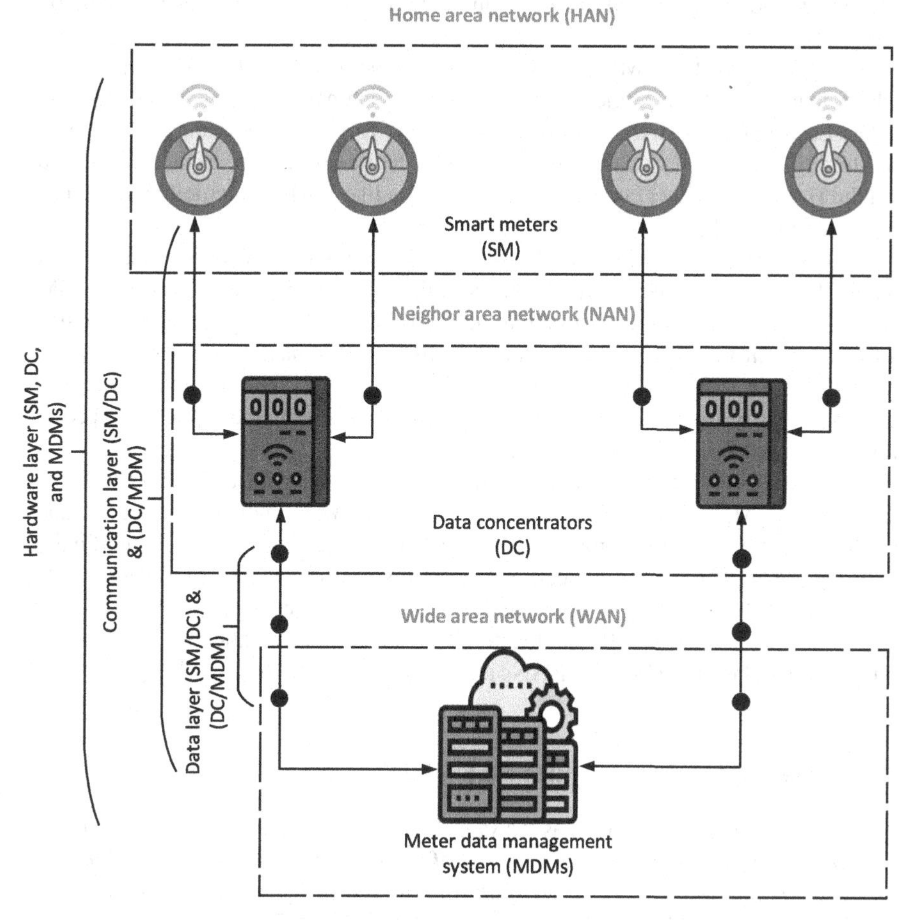

Figure 1.2 AMI components including the AMI layers and tiers. The SM, DC, and MDMs represent the AMI hardware layer, the interconnections between these devices represent the AMI communication layer, and the data transferred represent the AMI data layer.

electricity consumption. They also include firmware that has been transferred to both SM and DC as well as MDM setups and commands sent from MDMs to SM or DC [23]. The communication layer, the final layer of the AMI system, consists of the communication pathways that run from the SM to the MDMs through the DC and can be either wired or wireless in design [23].

1.2.3 AMI TIERS

According to Figure 1.2, there are three main stages of the AMI. Each stage includes a few of the AMI system's parts. The home area network (HAN), neighbor area network (NAN), and wide area network (WAN) are the three stages of an AMI. The components in each stage of an AMI can be connected via various communication protocols [24].

The first tier of the AMI is the HAN and is positioned on the client's property, which may be either a home or a business establishment. It connects an SM and IEDs. Each IED in a HAN expends energy in the form of data, which is sent to an SM through the two-way transmission link. Due to the limited size of the data carried via the HAN, low-power communication methods like Wi-Fi, Bluetooth, and Zigbee can be employed [25].

As part of the AMI structure, the NAN is the second stage, which can connect many HANs. The primary part of a NAN is the DC, which receives data from SM via HANs. The communication system needs to be capable of transmitting massive amounts of information more securely because a NAN has more subscribers and more data than a HAN. Hence, the most widely employed communication technologies in the NAN are mobile networks (LTE/2G-3G systems), fiber optics, and power line communication (PLC) systems [26].

The WAN serves as the final tier, which links all the HANs and NANs with the UC. The MDM system, which is connected to the WAN, gathers, maintains, and analyzes the data to fetch commands or instructions directly to the NAN and finally onto the HAN. Wired communication is chosen since it can send the obtained data more effectively and securely due to the great path length between the NANs and the UC, which are located in different locations. Power line communications (PLCs) are widely utilized to link HANs to NANs and subsequently NANs to WAN [27].

1.2.4 INFORMATION SECURITY RISK ASSESSMENT

Critical infrastructure is defined as *"systems and assets, whether physical or virtual, so critical that the inability or destruction of such systems and assets would have a catastrophic impact on security, national economic security, public health or safety, or any combination of those things,"* as declared in the NIST framework for improving the cybersecurity of critical infrastructure [28].

The critical information infrastructure systems, according to the NIST framework, are those whose operations depend on technology in general, including information technology (IT), industrial control systems (ICS), and connected devices, more generally, such as the IoT. Because of the reliance on technology, communication, and connectivity, possible vulnerabilities have changed and grown, and the risk to operations has increased [29].

As evidenced by its major aim, the AMI system can be regarded as one of the currently important systems whose operation is dependent on data gathered from IoT devices represented by SM and DC. Due to the importance of the information it delivers, the AMI system functions as one of the fundamental information infrastructure systems [30]. Due to the AMI system's potential flaws, attacks from both internal and external sources could compromise these data and result in electricity theft, customer data theft, partial blackouts, and large outages [31]. As a result, it's imperative to assess the AMI system in the context of the possible risk.

Information security risk assessment (ISRA) is the cornerstone of the risk management process. ISRA aims to identify the existing critical assets in the organization and their valuation, assess the existing vulnerabilities, and identify the probability of attacks occurring via various threat agents and their consequences on both the crucial assets and the overall system functions [32]. Finally, the existing security controls must be assessed to determine if they are effective and adequate or if extra security controls are required [33]. ISRA comprises three main tasks that are determined as follows: risk identification, risk profiling, and risk treatment, as shown in Figure 1.3 [34].

The first task of ISRA is risk identification, which establishes the main organization's assets, vulnerabilities, threats, asset valuations, and criticality level. The critical assets of any critical system may include people, processes, technologies, and information, regardless of its state of processing, storage, or transmission. Also, the possible threat agents that can exploit existing vulnerabilities must be considered during this step [32].

The following task of ISRA, known as risk profiling, combines the existing assets, vulnerabilities, threats, and possible risks with the risk score value to simplify the evaluation of risk criticality. This stage establishes the current security controls and, using an evaluation procedure, ranks the current threats. Choosing whether a quantitative or qualitative risk assessment approach will be employed is one of the stage's primary goals [32].

Risk treatment is the last phase of the ISRA task. The organization must choose the risk mitigation strategy to bring the risk's impact to a level that is manageable during the risk treatment stage. Risk mitigation can be accomplished by either sharing the risk with other third-party entities or adding some countermeasure techniques to lessen the consequences of the risks. Additionally, the organization may choose to

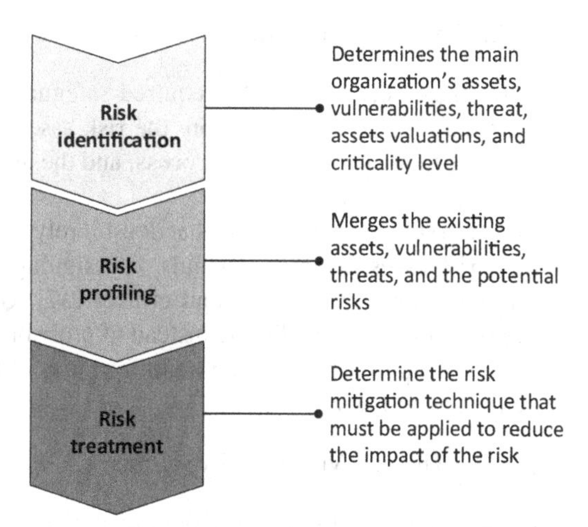

Figure 1.3 Information risk assessment's main stages: risk identification, risk profiling, and risk mitigation.

accept the current risk if it is within its risk-tolerance parameters and thus agrees to bear the cost when it arises [32].

Risk analysis is the process of assessing both the possibility that the threat will arise and the side effect of those dangers on the targeted system, which in our circumstance is the AMI system. It is an essential phase in the ISRA, especially with a large quantity of risks, which helps to prioritize risks and choose the proper risk mitigation technique for each risk. One of ISRA's primary responsibilities is determining the suitable risk analysis approach that can be carried out by each entity, depending on the system that needs to be evaluated and the capabilities of each risk analysis approach. There are multiple risk analysis approaches, such as quantitative, qualitative, semi-quantitative, knowledge-based, model-based, questionnaire, and expert systems [13].

Quantitative risk analysis is an objective approach that is based on a set of formulas, rules, and equations that focused on using the monetary value of the critical assets in the system. It can be applied mostly when the cost-benefit analysis (CBA) of the risk assessment process is required [35]. There are three main calculations that are applied in the quantitative approach: single loss expectancy (SLE), the annualized rate of occurrence (ARO), the annualized loss expectancy (ALE), and the safeguard value, as shown in the following formulas [36]:

$$\text{ALE} = \text{SLE} \times \text{ARO} \tag{1.1}$$

Equation 1.1 shows how to calculate the annualized loss expectancy of the asset.

$$\text{SLE} = \text{AV} \times \text{EF} \tag{1.2}$$

Equation 1.2 illustrates the calculation formula of single loss expectancy based on the asset value (AV) and exposure factor (EF).

$$\text{Safeguard value} = (\text{ALE before} - \text{ALE after}) - \text{Annual safeguard cost} \tag{1.3}$$

Equation 1.3 demonstrates the calculation of the required safeguard value for the asset based on the annualized loss expectancy before the risk assessment task, annualized loss expectancy after the risk assessment process, and the annual safeguard cost.

Qualitative risk analysis is a subjective approach that doesn't rely on the numerical values as with quantitative risk assessment but depends on assigning relative values for the security risk variable such as low, high, and critical [37]. Qualitative risk analysis approach uses the expression of likelihood instead of probability that is used in the quantitative risk analysis approach [38]. The main formulas of the qualitative risk assessment approach are as follows:

$$\text{Risk} = \text{Threat} \times \text{Vulnerability} \times \text{Impact} \tag{1.4}$$

Equation 1.6 shows the calculation of the risk for the qualitative risk analysis approaches based on the existing vulnerabilities, potential threats, and the overall consequence.

$$\text{Likelihood} = \text{Threat} \times \text{Vulnerability} \tag{1.5}$$

Equation 1.5 illustrates the relation between the likelihood of threats' occurrence, the current weaknesses, and the potential threat for the qualitative risk analysis approach.

Semi-quantitative represents a combination of both quantitative and qualitative risk assessment approaches to get the benefits from both. It has the advantages of simplicity as a qualitative approach and high accuracy level as a quantitative approach [33]. Its principle of operation depends on using approximated ranges (e.g., 0–15, 16–35,...) and then translating these ranges to qualitative values to simplify the comparison between the values within the same range or in different ranges [39].

A model-based approach is another example of the risk analysis approaches. One of the most prominent example of the model-based risk assessment approach is the unified modeling language (UML) that requires both a tool and an expert to perform the risk assessment procedures [32]. The principle of operation for the model-based risk assessment approach is creating at least five basic diagrams that are represented by the assets diagram, threat diagram, risk diagram, treatment diagram, and treatment overview diagram [40].

1.3 IMPLEMENTATION OF THE AMI SYSTEM'S RISK ASSESSMENT

The AMI system provides advanced features than the traditional electricity grid as energy monitoring, recording, data collection, load management, and advanced control capabilities. These features improve the efficiency of the grid and thus enable the customers to manage their electricity usage easily. The major objectives of AMI system are collecting the data from the end customer, transferring the data to the ESP for further analysis, and sending the control command from ESP to the end devices such as DC and SM for remotely managing, updating firmware, or remotely shutting down [41].

Considering the significance of the AMI system's role and its high criticality level, the task of risk assessment has to be accomplished to the AMI system to define the existing vulnerabilities, the potential threats, and the consequence of these threats on the AMI performance. The appropriate mitigation technique will then be employed to reduce the risk to an adequate level for the regular functionality of the AMI system.

1.3.1 RISK IDENTIFICATION PHASE FOR THE AMI SYSTEM

As shown in Figure 1.3, the initial stage of the risk assessment procedure on any system is the risk identification step that focuses on determining the main critical assets in the system that is required to be assessed and considered as the major step in any risk assessment process. Thus, the primary goal of this section is to identify the key and essential components of the AMI system, which will be used further in determining the main requirements for making use of the risk assessment procedure to AMI system.

The objects that need to be safeguarded, are likely to be exposed, and have a negative impact on the overall effectiveness and intent of the AMI are referred to as the

AMI assets. Any item that has the potential to affect the three key security properties, namely, confidentiality, integrity, and availability, if a threat takes advantage of any flaws in it can be considered an asset of the AMI [42].

AMI assets can be divided into tangible and intangible assets. AMI tangible assets start from the end customer in the entity, which are home or industry, passing by the SMs, DCs, and the MDMs in the data center of the ESP [43].

The data sent across communication lines or present on the SMs, DCs, and MDMs, which can be represented by their firmware, are another essential asset of the AMI. AMI intangible assets can be represented by the ESP's reputation that may be affected by the availability and confidentiality security attributes [44].

AMI assets can be classified into three main categories, namely, information assets, AMI resource assets, and services assets, as shown in Table 1.1. Information assets for the AMI can be determined as the audit data, the information about the energy usage for the customer, any policies or configurations in the MDMs of the ESP, locally protected information in the main data center, and the data that are transmitted through HAN, NAN, or WAN [24].

The resource assets of the AMI can be illustrated through its core components: SM, DC, and communication links. Additionally, the software and applications installed within the ESP data center are taken into account. Furthermore, other hardware within the AMI system is encompassed as smart appliances from the customer's perspective, and any hardware located in the primary data center. Finally, if a token is employed in the AMI system to verify the customer's validity, it will be added to the AMI resource assets [45].

The public key infrastructure (PKI) service, that can be employed as an authentication method for the essential AMI elements, may be one of the service assets. Upgrading and configuring the firmware of the SM or DC remotely are AMI services. The initialization step that is performed for the SM can be included as an AMI service asset. Furthermore, the access control service whether for the MDMs, which are located in the main data center, or for protecting the end devices from unauthorized access can be included as one of the AMI service assets. [46].

Table 1.1
The AMI Asset Classifications

AMI Resources Assets	AMI Data Assets	AMI Services Assets
SM, DC, and comm. links	Audit data	PKI
MDMs' software	Energy consumption	Remotely config.
Smart appliances	MDMs's configurations	Access control
Token for authentication	Local data in ESP	CIA

The main services that are provided by the AMI include confidentiality, integrity, and accountability services that can be provided to secure the customer data. The system's accessibility, and the main security services that are applied by the main data center administrator can also be included as an AMI service asset [14].

1.3.2 AMI VULNERABILITIES

The first process of the risk assessment also includes determining the existing vulnerabilities of the AMI system. Security is a new concern for this AMI system due to the integrating ICT into conventional power systems. The majority of the AMI components, including the SMs and DCs, are located outside the UC, making them susceptible to both physical and digital attacks. This section will cover security concerns pertaining to the essential hardware and software elements of an AMI. We will think about weaknesses, potential attacks, and how an attack would affect the components [47].

1.3.2.1 AMI Hardware Layer Vulnerabilities

As depicted in Figure 1.2, the SMs and DCs make up the AMI system's hardware layer. An SM outside of the UC has a number of risks due to its nature. Additionally, as we explained previously, a UC can command an SM by sending commands across a bidirectional channel to the device. The remote disconnect command can be used by the UC to start up or shut down the SM. This characteristic is necessary, but it also leaves the SM vulnerable to exploitation from someone pretending to be the UC. This command can be sent by the attacker to numerous SMs, which could result in a denial-of-service (DoS) attack [45].

Due to an SM's limitation of resources, such as internal storage, another security issue develops. A surprisingly small amount of storage is available on an SM, and it is utilized for both the firmware and the data gathered from users. Due to its limited storage capacity, firmware upgrades that are crucial, particularly security updates, may not be possible. Additionally, the SM could not have sufficient storage for encryption operations. They could be kept outside the SM chip on supplementary disc storage, in which they're exposed to traditional and cyber incursion. An attacker may take advantage of this weakness by using a buffer overflow attack to send out harmful data that quickly drain the SM buffer and make it stop accepting data [42].

A significant aspect of an SM is that customers may open it through a web page, enabling them to track energy usage and make payments. Furthermore, this functionality implies that data saved in the SM, such as information on energy usage or loaded firmware, are susceptible to intrusions via the web application, such as distributed denial of service (DDoS) attacks, and SQL injection attacks. An authentication bypass attack can be carried out by an attacker. This compromises the confidentiality, privacy, and accessibility of the consumer's personal data because they can alter or steal data from the SM even without knowing the customer's login information [43].

These flaws and cyberattacks that target an SM may have a detrimental effect on the functionality and performance of the AMI system. One of the main outcome of an attack on an SM is the stealing of the software installed on the SM or information about energy utilization. Understanding the energy usage of the client can expose the particular features of the IED they use, putting their privacy at danger. Additionally, a change to the firmware may cause the SM to malfunction and become inoperable [48].

Attacks similar to those against SM can also be made to a DC because numerous SMs are connected to the UC by a single DC, but they may have a higher impact, particularly if the AMI has either an indirect or a mesh structure architecture. A DC attack can have an impact on several SMs, resulting in the localized denial of power for multiple SMs at once or the theft of data from all the consumers linked to the hacked DC [49]. Table 1.2 shows the weaknesses including both SMs and DCs, the cyberattacks that can be launched to exploit them, and the effects of these threats on either the target components of the AMI or the entire AMI system.

1.3.2.2 AMI Data Layer Vulnerabilities

According to what had been discussed earlier, the AMI's major job is to gather information from SMs and send it by DCs toward the UC, to be stored and processed. The AMI design facilitates the transport of enormous amounts of data. Components of the AMI's status, such as security logs, the users' personal information, and energy consumption, are all included in these data. These data are susceptible to attacks including data insertion, data modification, and data hijacking, as listed in Table 1.3. Any of the three AMI layers can be used to carry out these threats [50].

Hackers might update or steal data and jeopardize the service's security or privacy if they were able to modify or insert data into the HAN. The WAN's data, which is located in the UC, could be modified by an internal or external attacker. For instance,

Table 1.2
Attacks That Exploit Security Flaws in the AMI Hardware Layer

AMI Item	Vulnerabilities	Attack	Consequence
SM & DC	Bidirectional communications channels	Sends malicious code to SMs	Denial of power
SM	Inadequate resources	Buffer overflow	Local denial of power
DC	DC outside the UC of the DC	Modify the firmware	Data theft
MDMs	Improper authentication configuration	Steal the MDMs configurations	Widespread denial of power

Table 1.3
AMI Data Layer Vulnerabilities and Related Exploits

Vulnerabilities	Attack	Impact
Customer and the SM direct connection	Modify energy consumption	Data loss or alteration
Remotely updating feature	Firmware manipulation	SM and DC shut down
Customer to SM interference	Fake data injection	Electricity fraud
Lack of security configuration in the UC	MDMs configuration manipulation	compromise the integrity of the data

they might alter the instructions given to the SMs, which would make the SG unstable and perhaps result in a denial of power.

The fact that an AMI uses the internet protocol (IP) to send data renders it susceptible to attacks that are successful targeting systems based on IP, such as teardrop attack, IP spoofing, and others that can result in data loss [51].

Table 1.3 outlines the relationships for both the AMI system's data layer vulnerabilities, related assaults, and the effects of those incidents. Because of the close relationship between the user and the SM, data alteration, firmware modification, and illegitimate tampering with a user's electricity utilization are the three categories of attacks against an AMI system's data layer currently.

1.3.2.3 AMI Communication Layer Vulnerabilities

As was previously mentioned, IEDs are coupled to SMs, SMs with DCs, and DCs with UC via communication links. Thus, they are the component of the AMI system that is most susceptible to dangers because of the significance and volume of data sent via the means of the channels of communication. HANs and NANs utilize wireless communication technology, making them vulnerable to man-in-the-middle (MITM) attacks and other cyberattacks on wireless communication technologies. This could result in the loss of consumer information [52].

One more weakness of the communication links is the breakdown of a communication channel, which might transpire due to disturbance, cable failures, network degradation, or degradation of channel capacity. This could result in isolated, widespread power outages and endangering the service's accessibility [53].

Table 1.4 outlines the relationships between the AMI communication layer's vulnerabilities, the attacks that take advantage of them, and the effects on the system's performance and other characteristics. The usage of wireless connectivity between the SMs and DCs is the primary flaw in the AMI system's communication layer. Wireless transmissions are susceptible to widespread assaults such as failure of the communication links and session hijacking that can result in data loss or fraud.

Table 1.4
Security Flaws in the AMI Communication Layer and Subsequent Threats

Vulnerabilities	Attack	Impact
Wireless communication cation technology used	Man-in-the-middle attack	Data theft
Wireless communication cation technology used	Session hijacking	Data manipulation
Wireless communication technology used	Bandwidth congestion	latency in data transfer

1.3.3 RISK PROFILING PHASE FOR THE AMI SYSTEM

After completing the first step of the risk assessment process on the AMI system, which focused on identifying the system's primary assets, vulnerabilities, and potential points of attack, risk profiling, as shown in Figure 1.3, is the next stage of evaluating the security risk associated with the AMI system. This stage focuses on matching the predefined AMI system's critical assets, the attack, and the potential risk.

By assessing the possibility of the attacks occurring and their effect on the AMI system, the goal of this stage is accomplished. Additionally, the likelihood level and the consequence level are matched to create the risk matrix, which is then used to rank probable risks. The risk matrix will be utilized in the final step of the risk assessment procedure to choose the best risk management strategy, which will be followed by determining the possible risk level associated with each attack.

Identifying the possibility of an attack on the AMI system occurring is the initial step in this phase. The likelihood of the attack is mostly determined by the same factors as listed below: the attacker's motivation for attacking the AMI system, the threat agent's capabilities, and the potential attack opportunities [54].

There are a number of reasons why threat agents might target the AMI system, both internally and externally. Avoiding billing and electricity theft from the customer side is one of the well-known motives. External attackers are motivated for a wide range of causes including the points outlined below: to reduce revenue, harm a company's reputation, prevent the deployment of meter, harm infrastructure, compromise confidentiality, and manipulate the energy market [54].

For a threat agent to successfully attack the AMI system, they need to acquire a certain set of skills. The cost, the attacker's skills, the attack's execution time, and the vulnerabilities already in place that the attacker will use are all included in the threat agent's capabilities [54].

The opportunity that the threat agent will exploit is the final factor in calculating the likelihood level. These chances can be identified by the AMI system's current weaknesses. One of the vulnerabilities that the attacker may exploit to compromise the MDMs is the absence of an access control mechanism [54].

The possibility of the attack occurring has four levels, as stated in Table 1.5, and they are *rare, possible, likely, and certain*. Rare likelihood level refers to the low probability that these attacks will occur. Possible level denotes that the attack could happen at any moment. Likely level indicates that these attacks will be probably performed in most circumstances. Certain level denotes that these attacks are typically predicted to occur [55].

The level of likelihood is dependent on the level of the purpose, means, and opportunity, as shown in Table 1.5. It is demonstrated that the likelihood value is directly proportional to the purpose, mean, and opportunity levels. The possibility of the attack occurring is reduced because the motive level is low. The likelihood of an attack occurring decreases with decreasing threat agent capabilities. The attacker has few opportunities, which reduces the possibility of an attack.

The level of consequences is the next step in this phase. The attack's potential impact on the AMI system's essential components and function is indicated by the attack's consequence level. The three effect levels, which are negligible, substantial, and catastrophic, are shown in Table 1.6. Insignificant consequence levels signify that only one AMI system component is compromised, and as a result, they do not have an impact on the entire AMI system. The attack did not disrupt the main operations of the AMI system, but various AMI components were affected, according to the

Table 1.5
Likelihood Levels of the Attack Occurrence

Motive	Mean	Opportunity	Likelihood
Low	Low	Low	Rare
Low	Low	High	Possible
Low	High	Low	Possible
Low	High	High	Likely
High	Low	High	Likely
High	High	Low	Likely
High	High	High	Certain

Table 1.6
Consequence Levels of the Attack

Consequence Level	Definition
Insignificant	The AMI function is slightly impacted
Moderate	The AMI function is degraded significantly
Catastrophic	Inflicting the AMI system tremendous damage

Likelihood		Consequences		
		Insignificant	Moderate	Catastrophic
	Certain	M	E	E
	Likely	M	H	E
	Possible	L	M	H
	Rare	L	M	H

Figure 1.4 Risk matrix: matching between the likelihood and consequence levels.

moderate consequence level. The catastrophic level denotes that the attack affected every AMI system's components, impairing all of its operations and may cause a widespread blackout [56].

The following step in this phase is matching the likelihood and consequence levels that were previously computed to obtain the risk matrix. The risk level depends on both the likelihood and consequence levels, as illustrated in Eq. 1.6 [57].

$$\text{Risk} = \text{Likelihood} \times \text{Consequence} \tag{1.6}$$

The risk matrix, which displays the risk levels based on both the likelihood and consequence levels, is shown in Figure 1.4. There are four risk categories, namely, low risk (L), moderate risk (M), high risk (H), and severe risk (E). The risk can be accepted, according to the low-risk value. The moderate-risk level denotes the necessity of taking steps to monitor the risk. An action must be taken to reduce the danger level due to the high-risk level. The extreme-risk level signifies that the response will be highly appreciated and is necessary to reduce the risk.

This phase's final step involves assessing the predetermined critical assets, vulnerabilities, assaults, likelihood, consequence, and risk levels to create risk scenarios. These risk scenarios will include the hardware, data, and communication layers of the AMI system, all of which were previously specified.

AMI system attack scenarios, the AMI layer that is impacted, the possibility that the attack would occur, the severity of each attack scenario, and the risk involved are shown in Table 1.7. Based on the risk matrix displayed in Figure 1.4, the risk level is computed.

1.3.4 RISK TREATMENT PHASE FOR THE AMI SYSTEM

The AMI system's critical components, weaknesses, and potential assaults are first identified. In the preceding part, the likelihood, impact, and risk matrix were also established. Finally, several attack scenarios specific to the AMI system were outlined. The risk treatment phase, which will be our next phase, is the final stage of the risk assessment, as shown in Figure 1.3. The risk treatment phase focuses on selecting the best approach to reduce the unacceptable risk to a manageable level. There are three methods for dealing with risks, namely, accepting the risk, reducing the risk, and transferring the risk [58].

By implementing security measures like encryption, authentication, and intrusion prevention systems, the risk can be reduced. These security measures help to bring

Table 1.7
Risk Scenarios: Including the Assets, the Likelihood, and Consequence Levels

Risk Code	AMI Layer	Attack Scenario	Likelihood Level	Consequnce Level	Risk Level
R1	Data	Breaks the applied encryption	Rare	Catastrophic	H
R2	Data	Firmware manipulation	Likely	Moderate	H
R3	Hardware	Injecting malicious code	Likely	Moderate	H
R4	Communication	Eavesdrops wireless communication channel	Likely	Moderate	H
R5	Data	Eavesdrops to steal the data	Likely	Moderate	H
R6	Hardware	Gains improper access to assets	Likely	Moderate	H
R7	Hardware	Discovers MDMs admin authentication information	Likely	Catastrophic	E
R8	Data	Access browses files to collect information	Likely	Moderate	H
R9	Data	Duplicate keys of the PKI system	Rare	Moderate	H

down the danger to a manageable level. If the risk has little impact on how the system functions, it can be accepted, and it is preferable to do so than accepting the financial burden of adding more safeguards. Risks may be transmitted to any other third party for treatment [58].

The suitable risk treatment approach will be chosen using the risk matrix, as indicated in Figure 1.4. As shown in Figure 1.4, the *L letter* denotes a risk that can be accepted, whereas the *M*, *H*, and *E* letters represent risks required to be reduced by implementing extra security controls.

1.4 DISCUSSION AND RECOMMENDATIONS

By assessing the AMI system against the potential cybersecurity risks, there are nine risk scenarios associated with the AMI system, as shown in Table1.7. Each risk scenario has been linked to its relevant AMI layer. The likelihood, consequence, and risk level of each risk scenario are also determined. It can be determined that certain attack scenarios such as R1, R2, R4, R5, R6, R7, and R9 have high-risk levels, whereas R8 has an extremely high-risk level.

Using the risk matrix provided in Figure 1.4, the attack scenarios, the risk level for each risk scenario, and the risk management techniques are shown in Table 1.7. All of these risk scenarios must be reduced.

It is possible to secure the AMI system from cyberattacks and minimize the current risk levels to acceptable levels by using a variety of mitigation strategies. The three major risk mitigation tactics are intrusion detection system (IDS), authentication, and encryption. These mitigation strategies will be discussed in this section.

Encryption is a crucial method for safeguarding the AMI system's data layer. The risk scenarios relating to the AMI's data layer can be mitigated with encryption technique. By encrypting the data passing through the AMI system using a strong encryption approach, the risk scenarios R1, R2, R5, and R8 can be reduced.

Preserving the data conveyed via an AMI system requires confirming the data's origins. Procedures for authentication are essential to safeguard the hardware layer of the AMI, covering SMs and DCs. Applying a robust authentication technique to the AMI system will minimize the security risk scenarios R3, R6, and R7 that are previously determined.

Creating node-to-node network interactions requires authentication. Authentication can stop attacks when an attacker pretends to be a UC and transmits shutdown notifications to several SMs. Additionally, the attacker can transmit harmful data to the DC, potentially affecting the DC's firmware and the DC's performance, thus the performance of the entire SG [59].

An IDS is a potent tool for protecting an AMI system's data and communication layers from attackers who may otherwise take advantage of undiscovered system gaps and cause disruptions. IDSs are regarded as a system's initial line of defense. Clients can use application server to offer up information or make payments. Therefore, a privacy protection step is needed to secure the confidentiality of this critical data [60].

1.4.1 RECOMMENDATIONS

Fully homomorphic encryption (FHE) is a type of encryption that relies on running calculations on the encrypted data without first decrypting it. According to Ref. [61], these techniques can be applied to systems that include sensitive data, like the AMI system. A distributed ledger that can be used for encryption is known as a blockchain. As a result, a blockchain is a potent new tool for cybersecurity since it can foster trust in an untrustworthy environment. The decentralized ledger includes transactions from numerous devices. To guarantee data integrity, the blockchain's participants encrypt and store the data [62].

The AMI system can benefit from the usage of cloud computing technologies as authentication mitigation method. The cloud computing technology clusters system components into security groups as a default isolation feature [63]. This helps to stop disallowed access or the installation of harmful data to the end equipment by attackers since the possibilities of interaction between entities by default are limited. The authentication and authorization methods necessary for the AMI system can be improved with the help of this feature.

Artificial intelligence (AI) is employable for securing an AMI system as a mitigation strategy. An AI system may gradually understand the behavior of every device in the AMI system via deep learning and machine learning. As a result, an AMI system may discover and identify dangerous behavior faster thanks to AI than it can with conventional methods. Because of its automatic ability to scan through enormous volumes of both data and traffic, AI is able to discover dangers that are concealed as a routine activity [64].

AI has the ability to quickly scan a whole AMI system for vulnerabilities, both known and undiscovered, that might one day be exploited. Many IDSs that are used to find vulnerabilities and threats use AI at their core. AI can enable reliable client verification utilizing biometrics like fingerprint readers and facial recognition [65].

1.5 CONCLUSION

One of the existing examples of an IoT system is the AMI system, which uses IT to digitize the conventional electricity grid. The primary goal of the AMI system is to gather data from the customer domain utilizing the SM, DC, and UC, so that it may be gathered, stored, and analyzed by the MDMs. These data are traveling through the AMI system via bidirectional communication channels.

The spread of interconnected devices presents a variety of security challenges for IoT technology, increasing the potential for attacks. The IoT system faces a security difficulty due to the linked devices' non-traditional locations, which makes achieving physical security challenging or impossible. IoT sensor-enabled devices may be situated in remote or inhospitable areas, making it nearly difficult for humans to configure or intervene. Thus, the likelihood of upgrading these devices, which leave them susceptible to attacks, is diminished.

One of the approaches that should be used to evaluate the effectiveness of the security controls currently in place for any important system, including the AMI system, is risk assessment. The outcome of using the ISRA on the AMI system may help in preventing any anticipated attacks on the system.

The three primary ISRA phases were applied to the AMI system in this chapter. When employing the ISRA to the AMI system, the critical assets, vulnerabilities, and potential attacks that can be conducted against each component of the AMI system are shown. The risk matrix and the possibility of the occurrence, impact, and attacks were shown. Additionally, nine risk scenarios are obtained and linked with the AMI layer that was affected by the attack, the chance of its occurrence, and the severity of its effects, and the risk level of each attack scenario was displayed.

Finally, a suitable risk treatment action that can be accepted, mitigated, or transferred must be applied depending on the risk level for each attack scenario. This chapter provided examples of countermeasure methods that might be used to strengthen the AMI system's security safeguards. Encryption, authentication, and IDS are the mitigation strategies that have been suggested, which can be combined with the AMI system to increase system security.

ACKNOWLEDGMENT

This study was supported by a joint United Arab Emirates University and Zayed University (UAEU-ZU) research grant (Grant number: 12R141).

REFERENCES

1. Mrunal M. Kapse, Nilofar R. Patel, Shruti K. Narayankar, Sachin A. Malvekar, and Kazi Kutubuddin Sayyad Liyakat. Smart grid technology. *International Journal of Information Technology & Computer Engineering (IJITC)*, 2(06):10–17, 2022.
2. Priyanka D. Halle and Subramani Shiyamala. Secure advance metering infrastructure protocol for smart grid power system enabled by the internet of things. *Microprocessors and Microsystems*, 95:104708, 2022.
3. Bhabendu Kumar Mohanta, Debasish Jena, Utkalika Satapathy, and Srikanta Patnaik. Survey on IoT security: Challenges and solution using machine learning, artificial intelligence and blockchain technology. *Internet of Things*, 11:100227, 2020.
4. Sachchidanand Singh and Nirmala Singh. Internet of things (IoT): Security challenges, business opportunities & reference architecture for e-commerce. In *2015 International Conference on Green Computing and Internet of Things (ICGCIoT)*, Greater Noida, India, pp. 1577–1581. IEEE, 2015.
5. Lo'ai Tawalbeh, Fadi Muheidat, Mais Tawalbeh, and Muhannad Quwaider. IoT privacy and security: Challenges and solutions. *Applied Sciences*, 10(12):4102, 2020.
6. Mohsin Kamal and Muhammad Tariq. Light-weight security and blockchain based provenance for advanced metering infrastructure. *IEEE Access*, 7:87345–87356, 2019.
7. Mostafa Shokry, Ali Ismail Awad, Mahmoud Khaled Abd-Ellah, and Ashraf A. M. Khalaf. Systematic survey of advanced metering infrastructure security: Vulnerabilities, attacks, countermeasures, and future vision. *Future Generation Computer Systems*, 136(1):358–377, 2022.
8. Josė Luis Gallardo, Mohamed A. Ahmed, and Nicolás Jara. Lora IoT-based architecture for advanced metering infrastructure in residential smart grid. *IEEE Access*, 9:124295–124312, 2021.
9. Arrizky Ayu Faradila Purnama, and Muhammad Imam Nashiruddin. Designing lorawan internet of things network for advanced metering infrastructure (AMI) in surabaya and its surrounding cities. In *2019 International Seminar on Research of Information Technology and Intelligent Systems (ISRITI)*, Yogyakarta, Indonesia, pp. 194–199. IEEE, 2019.
10. Defense Use Case. Analysis of the cyber attack on the ukrainian power grid. *Electricity Information Sharing and Analysis Center (E-ISAC)*, 388, 2016.
11. Andrea Fried, Svjetlana Pantic Dragisic, Arne Jönsson, and Mona Mirtsch. Communicating preventive innovation-the case of the information security standard ISO/IEC 27001. In *European Group of Organization Studies Colloquium 2022, Subtheme 6 on Performing Creativity, Innovation, and Change: Communicating to Reconfigure the Organization, Wirtschaftsunversität Wien*, Austria, 2022.
12. Hasan Mahbub Tusher, Ziaul Haque Munim, Theo E. Notteboom, Tae-Eun Kim, and Salman Nazir. Cyber security risk assessment in autonomous shipping. *Maritime Economics & Logistics*, 24(2):208–227, 2022.
13. Alireza Shameli-Sendi, Rouzbeh Aghababaei-Barzegar, and Mohamed Cheriet. Taxonomy of information security risk assessment (ISRA). *Computers & Security*, 57:14–30, 2016.
14. Mungyu Bae, Kangho Kim, and Hwangnam Kim. Preserving privacy and efficiency in data communication and aggregation for AMI network. *Journal of Network and Computer Applications*, 59:333–344, 2016.

15. Otisitswe Kebotogetse, Ravi Samikannu, and Abid Yahya. Review of key management techniques for advanced metering infrastructure. *International Journal of Distributed Sensor Networks*, 17(8):15501477211041541, 2021.
16. Alireza Ghasempour. Optimized advanced metering infrastructure architecture of smart grid based on total cost, energy, and delay. In *2016 IEEE Power Energy Society Innovative Smart Grid Technologies Conference (ISGT)*, Minneapolis, MN, USA, pp. 1–6, 2016.
17. Vincenzo Gulisano, Magnus Almgren, and Marina Papatriantafilou. Metis: A two-tier intrusion detection system for advanced metering infrastructures. In *International Conference on Security and Privacy in Communication Networks*, Cambridge, United Kingdom, pp. 51–68. Springer, 2014.
18. Neetesh Saxena, Bong Jun Choi, and Rongxing Lu. Authentication and authorization scheme for various user roles and devices in smart grid. *IEEE Transactions on Information Forensics and Security*, 11(5):907–921, 2015.
19. Alireza Ghasempour. *Optimizing the Advanced Metering Infrastructure Architecture in Smart Grid*. Utah State University, 2016.
20. Asad Khattak, Salam Khanji, and Wajahat Khan. *Smart Meter Security: Vulnerabilities, Threat Impacts, and Countermeasures*, pp. 554–562, 2019.
21. Sandeep Kumar Singh, Ranjan Bose, and Anupam Joshi. Entropy-based electricity theft detection in AMI network. *IET Cyber-Physical Systems: Theory & Applications*, 3(2):99–105, 2018.
22. Ziad Ismail, Jean Leneutre, David Bateman, and Lin Chen. A game theoretical analysis of data confidentiality attacks on smart-grid AMI. *IEEE Journal on Selected Areas in Communications*, 32(7):1486–1499, 2014.
23. Kiyana Pedramnia and Masoomeh Rahmani. Survey of DoS attacks on LTE infrastructure used in AMI system and countermeasures. In *2018 Smart Grid Conference (SGC)*, Sanandaj, Iran, pp. 1–6. IEEE, 2018.
24. Rebeca P. Díaz Redondo, Ana Fernández-Vilas, and Gabriel Fernández dos Reis. Security aspects in smart meters: Analysis and prevention. *Sensors*, 20(14):3977, 2020.
25. Ye Yan, Rose Qingyang Hu, Sajal K. Das, Hamid Sharif, and Yi Qian. An efficient security protocol for advanced metering infrastructure in smart grid. *IEEE Network*, 27(4):64–71, 2013.
26. Ivan Popović, Aleksandar Rakić, and Ivan D Petruševski. Multi-agent real-time advanced metering infrastructure based on fog computing. *Energies*, 15(1):373, 2022.
27. Lili Yan, Yan Chang, and Shibin Zhang. A lightweight authentication and key agreement scheme for smart grid. *International Journal of Distributed Sensor Networks*, 13(2):1550147717694173, 2017.
28. Lei Shen. The NIST cybersecurity framework: Overview and potential impacts. *Scitech Lawyer*, 10(4):16, 2014.
29. Hermawan Setiawan, Fandi Aditya Putra, and Anggi Rifa Pradana. Design of information security risk management using ISO/IEC 27005 and NIST SP 800-30 revision 1: A case study at communication data applications of XYZ institute. In *2017 International Conference on Information Technology Systems and Innovation (ICITSI)*, Bandung, Indonesia, pp. 251–256. IEEE, 2017.
30. Ievgeniia Kuzminykh, Bogdan Ghita, Volodymyr Sokolov, and Taimur Bakhshi. Information security risk assessment. *Encyclopedia*, 1(3):602–617, 2021.
31. Vivek Agrawal. A comparative study on information security risk analysis methods. *Journal of Computers*, 12(1):57–67, 2017.

32. Gaute Wangen, Christoffer Hallstensen, and Einar Snekkenes. A framework for estimating information security risk assessment method completeness. *International Journal of Information Security*, 17(6):681–699, 2018.
33. Palaniappan Shamala, Rabiah Ahmad, and Mariana Yusoff. A conceptual framework of info structure for information security risk assessment (ISRA). *Journal of Information Security and Applications*, 18(1):45–52, 2013.
34. John R. S. Fraser, Rob Quail, and Betty J. Simkins. Questions asked about enterprise risk management by risk practitioners. *Business Horizons*, 65(3):251–260, 2022.
35. Yulia Cherdantseva, Pete Burnap, Andrew Blyth, Peter Eden, Kevin Jones, Hugh Soulsby, and Kristan Stoddart. A review of cyber security risk assessment methods for SCADA systems. *Computers & Security*, 56:1–27, 2016.
36. Jason R. C. Nurse, Sadie Creese, and David De Roure. Security risk assessment in internet of things systems. *IT Professional*, 19(5):20–26, 2017.
37. Douglas Landoll. *The Security Risk Assessment Handbook: A Complete Guide for Performing Security Risk Assessments*. CRC Press, Boca Raton, FL, 2021.
38. Xiaorong Lyu, Yulong Ding, and Shuang-Hua Yang. Safety and security risk assessment in cyber-physical systems. *IET Cyber-Physical Systems: Theory & Applications*, 4(3):221–232, 2019.
39. Yang Wang, Jun Zheng, Chen Sun, and Srinivas Mukkamala. Quantitative security risk assessment of android permissions and applications. In *IFIP Annual Conference on Data and Applications Security and Privacy*, Berlin, Heidelberg, pp. 226–241. Springer, 2013.
40. Paul Baybutt. Issues for security risk assessment in the process industries. *Journal of Loss Prevention in the Process Industries*, 49:509–518, 2017.
41. Rong Jiang, Rongxing Lu, Ye Wang, Jun Luo, Changxiang Shen, and Xuemin Shen. Energy-theft detection issues for advanced metering infrastructure in smart grid. *Tsinghua Science and Technology*, 19(2):105–120, 2014.
42. Megan Milam and Ganesh Kumar Venayagamoorthy. Smart meter deployment: Us initiatives. In *ISGT 2014*, Washington, DC, pp. 1–5, 2014.
43. Ivan Petruševski, Miloš Živanović, Aleksandar Rakić, and Ivan Popović. Novel AMI architecture for real-time smart metering. In *2014 22nd Telecommunications Forum Telfor (TELFOR)*, pp. 664–667, 2014.
44. Syeda Pealy and Mohammad Abdul Matin. A survey on threats and countermeasures in smart meter. In *2020 IEEE International Conference on Communication, Networks and Satellite (ComNetSat)*, Batam, Indonesia, pp. 417–422, 2020.
45. Justine L. Pesesky. The vulnerabilities of the advanced metering infrastructure in the Smart Grid. PhD thesis, Utica College, 2016.
46. Kallisthenis I. Sgouras, Athina D. Birda, and Dimitris P. Labridis. Cyber attack impact on critical smart grid infrastructures. In *ISGT 2014*, Washington, DC, USA, pp. 1–5, 2014.
47. Siyuan Dong, Jun Cao, David Flynn, and Zhong Fan. Cybersecurity in smart local energy systems: Requirements, challenges, and standards. *Energy Informatics*, 5(1):1–30, 2022.
48. Ke Zhang, Zhi Hu, Yufei Zhan, Xiaofen Wang, and Keyi Guo. A smart grid AMI intrusion detection strategy based on extreme learning machine. *Energies*, 13(18):4907, 2020.
49. Ahmed S. Alfakeeh, Sarmadullah Khan, and Ali Hilal Al-Bayatti. A multi-user, single-authentication protocol for smart grid architectures. *Sensors*, 20(6):1581, 2020.
50. Wenye Wang and Zhuo Lu. Cyber security in the smart grid: Survey and challenges. *Computer Networks*, 57(5):1344–1371, 2013.

51. Dipanjan Das Roy and Dongwan Shin. Network intrusion detection in smart grids for imbalanced attack types using machine learning models. In *2019 International Conference on Information and Communication Technology Convergence (ICTC)*, Jeju, Korea (South), pp. 576–581. IEEE, 2019.
52. Muhammad Qasim Ali and Ehab Al-Shaer. Randomization-based intrusion detection system for advanced metering infrastructure. *ACM Transactions on Information and System Security (TISSEC)*, 18(2):1–30, 2015.
53. Victoria Y. Pillitteri and Tanya L. Brewer. *Guidelines for Smart Grid Cybersecurity*. National Institute of Standards and Technology, Gaithersburg, MD, 2014.
54. Yonghong Li, Ruifeng Liu, Xiaoyu Liu, Hong Li, and Qingwen Sun. Research on information security risk analysis and prevention technology of network communication based on cloud computing algorithm. In *Journal of Physics: Conference Series*, Chongqing, China, vol. 1982, p. 012129. IOP Publishing, 2021.
55. Bunyamin Gunes, Gizem Kayisoglu, and Pelin Bolat. Cyber security risk assessment for seaports: A case study of a container port. *Computers & Security*, 103:102196, 2021.
56. He Li, Zhiang Zhang, Jia Sun, Ziyi Qi, Yue Min, and Zhi Qi. Smart grid information security evaluation method based on risk weight algorithm. In *International Conference on Algorithms, High Performance Computing, and Artificial Intelligence (AHPCAI 2021)*, Sanya, China, vol. 12156, pp. 213–217. SPIE, 2021.
57. Abhijit Guha, Debabrata Samanta, Amit Banerjee, and Daksh Agarwal. A deep learning model for information loss prevention from multi-page digital documents. *IEEE Access*, 9:80451–80465, 2021.
58. Vangelis Malamas, Fotis Chantzis, Thomas K. Dasaklis, George Stergiopoulos, Panayiotis Kotzanikolaou, and Christos Douligeris. Risk assessment methodologies for the internet of medical things: A survey and comparative appraisal. *IEEE Access*, 9:40049–40075, 2021.
59. Bakkiam David Deebak and AL-Turjman Fadi. Lightweight authentication for IoT/Cloud-based forensics in intelligent data computing. *Future Generation Computer Systems*, 116:406–425, 2021.
60. Ankit Thakkar and Ritika Lohiya. A review on machine learning and deep learning perspectives of IDS for IoT: Recent updates, security issues, and challenges. *Archives of Computational Methods in Engineering*, 28(4):3211–3243, 2021.
61. Samet Tonyali, Nico Saputro, and Kemal Akkaya. Assessing the feasibility of fully homomorphic encryption for smart grid AMI networks. In *2015 Seventh International Conference on Ubiquitous and Future Networks*, Sapporo, Japan, pp. 591–596. IEEE, 2015.
62. Michael Nofer, Peter Gomber, Oliver Hinz, and Dirk Schiereck. Blockchain. *Business & Information Systems Engineering*, 59(3):183–187, 2017.
63. Suyel Namasudra. Data access control in the cloud computing environment for bioinformatics. *International Journal of Applied Research in Bioinformatics (IJARB)*, 11(1):40–50, 2021.
64. Marcelo Zanetti, Edgard Jamhour, Marcelo Pellenz, Manoel Penna, Voldi Zambenedetti, and Ivan Chueiri. A tunable fraud detection system for advanced metering infrastructure using short-lived patterns. *IEEE Transactions on Smart Grid*, 10(1):830–840, 2019.
65. Kevin Song, Paul Kim, Vedant Tyagi, and Shivani Rajasekaran. Artificial immune system (AIS) based intrusion detection system (IDS) for smart grid advanced metering infrastructure (AMI) networks. 2018.

2 A Generative Neural Network for Improving Metamorphic Malware Detection in IoT Mobile Devices

Leigh Turnbull and Zhiyuan Tan
Edinburgh Napier University

Kehinde O. Babaagba
Edinburgh Napier University

2.1 INTRODUCTION

Android-based mobile devices contribute significantly to the evolution of IoT across several business sectors, including healthcare, industrial control systems, smart homes, and the automotive industry. However, due to Android's dominant position in the mobile market, the platform is the focus of regular targeted attacks [45]. Malicious code/software often referred to as malcode or malware is created to distort the functionality of a machine or network. These include several groups whose design functionalities differ such as Worms, Trojans, Adware, Rootkits, Spyware, Crypto malware, and Viruses. The capacity and complexity of these malware groups are ever growing particularly with technological advancement and the increasing awareness of vulnerabilities by attackers [28].

The design of advanced evasion techniques by malware attackers has been driven by improvements in malware detection. One of the foremost stealth tactics employed to evade detection was encryption, which made the malware undetectable before decryption. However, through string signature matching, the decryption module may be detected as malware. An advancement on encrypted malware is oligomorphism in which several randomly chosen decryption codes are used for each victim. A more complex type of an oligomorphic malware uses polymorphism wherein code encryption is employed with the added capability of creating an unlimited number of novel decryptors. Polymorphism is intended to continuously transform the decryption

 DOI: 10.1201/9781003199410-2

routine with each new instance and employs a mutation engine for the obfuscation of code using techniques such as garbage code insertion and instruction replacement to mutate the decryptor [6]. The main body of polymorphic malware code is consistent across new infections, and although it is not detectable in the encrypted form, it is detectable with signature matching once it is decrypted and loaded into memory, which is a functional requirement of polymorphic malware [59].

An evolution of polymorphic malware that does not rely on a decryption routine is metamorphic malware. This family of malware is regarded as a particularly dangerous malware group due to its ability to transform the main body of code at each new infection while preserving the underlying malicious functionality. As such, detecting metamorphic malware continues to pose a significant challenge due to its continuous code mutation functionality [59]. A more in-depth study of the behavior and anomaly analysis of metamorphic malware is deemed necessary to increase the possibility of detection [39].

Conventional signature-based approaches to malware detection fail to detect new mutants of malware and are evaded by these malware variants. More sophisticated methods of detecting malware now include advancements, such as using heuristics and machine learning, for detecting more and more dangerous malicious mutants. This includes techniques such as generative adversarial network (GAN), which was introduced as a probable solution to defeating metamorphic or zero-day malware attacks by creating novel samples and signatures evolved from existing malware [37]. In addition, the convolutional GAN framework was employed in the generation of adversarial examples that were able to evade third-party malware detectors comprising of techniques for the transformation of program process execution [68]. These techniques aimed at defecting complex malicious families also include techniques such as behavior profiling and analysis of malware, built on semantic tactics, in order to discover malicious patterns, providing awareness and insight into the functional operations of malicious mutants, and improve the discovery of obfuscation attempts made by such malware [63].

In this chapter, we lay the foundation for a practical understanding of the effectiveness of GAN in the improvement of metamorphic malware detection based on behavior profiling. The main research questions, for which this work will seek answers, include:

- What are the key features of metamorphic malware that may enhance detection using behavioral characteristics?
- How effective are GANs for enhancing the detection of metamorphic malware using behavioral characteristics?

The rest of the chapter is structured as follows. In Section 2.2 of the chapter, we provide a literature review. Section 2.3 describes our research methodology. The experimental design is given in Section 2.4. The results and discussion are provided in Section 2.5. Section 2.6 concludes the chapter and suggests areas of future research.

2.2 BACKGROUND

2.2.1 MACHINE LEARNING

Artificial intelligence (AI) is a subset of computer science aimed at developing technological advancements to enhance human intelligence and creating new intelligent technologies that mimic human intelligence. AI has facilitated innovation in computer vision, nature language processing, and robotics. Machine learning (ML) is a subset of AI which optimizes the use of mathematical processes to perform statistical analysis, classification, and prediction. ML is described as a "field of study that gives computers the ability to learn without being explicitly programmed" by AI pioneer Arthur Samuel [61].

The methodology adhered to in ML is performed in two stages, namely, a training phase and a testing phase. The stages typically involve actions to identify features and subset attributes from training data, algorithm selection, model training and evaluation, and final model selection for data classification. In practice, the ML procedure is performed by training, validation, and testing, and sample data are divided into training, testing, and validation datasets. Normal patterns are defined in the training stage using a training dataset, and the validation dataset is used to substantiate the efficacy of the process used in the training stage. The test dataset is used to verify the accuracy and overall efficiency of the model used [11].

ML algorithms may be categorized as deep or shallow learning and are further classified as supervised, unsupervised, or reinforcement learning systems. Deep learning involves more complexity and relies on interconnected networks to process input data via several intermediate layers to generate an output. Deep learning models are based on artificial neural networks (ANN) and can perform feature selection from input data to optimize performance and arbitrarily learn data. Vast amounts of complex data can be learned by deep learning algorithms that have been used for various purposes including malware analysis and threat identification [21]. Shallow learning requires manual feature identification and extraction and is dependent upon an understanding of the data being processed. Both approaches are further categorized as supervised learning, which requires classification and labeling of the training dataset, and unsupervised learning, which does not rely on a classified dataset to operate. Unsupervised learning is capable of independently identifying relevant features and categorizing data with related attributes, whereas supervised algorithms are commonly used for data classification based on feature correlation [3].

Shallow machine learning techniques for malware detection have been experimented with in the past, such as support vector machine, Naive Bayes, and decision tree algorithms, using various features. However, these techniques are not optimized for complex malware detection. Deep learning techniques mitigate some of the shortcomings of shallow learning due their enhanced feature learning capabilities and have demonstrated superior performance against shallow learning methods in past researches using stacked autoencoders [23]. The features required by machine learning classification algorithms are a key component of the training datasets, which impact the overall accuracy of machine learning. The extraction of pertinent features from training data is intended to optimize the process of differentiating between benign and malicious software. Feature extraction is challenging to achieve an optimal balance between speed, number of selected features, and accuracy [19].

The rest of this section lists the frequently extracted features in research for malware detection using ML. Dynamic link libraries (DLLs) function calls obtain information associated with Microsoft Windows DLLs and API functions, which may provide insight into the intended behavior of a program. This technique has been used to determine the DLLs in use by a binary file, API functions referenced, and number of API functions called by each DLL. Binary sequence extraction method has been researched using hexadecimal code, n-gram sequences, and fixed length byte values as features. Assembly sequences or opcode sequences have been used to identify and understand malicious functions from disassembled binary files, using n-gram sequences for feature extraction. Portable executable (PE) header fields contain structural information about binary files, which has been used for feature extraction. Integer and Boolean values derived from the PE header fields serve as feature options. Entropy signals are a measurement of randomness within data; high entropy within data indicates that there is greater randomness and content of information in a binary file, which may be represented as an entropy stream of code chunks that are developed into features. Machine activity metrics of the number of running processes, network activity, memory, and CPU usage have been used as features to indicate malicious process behavior [35].

API graph matching and similarity features have also been researched to detect malware based on the longest common subsequence algorithm that uses similarity measurements. The study demonstrated a 98% detection rate and 0% false positive; however, this study was limited to 75 malicious and 10 benign samples [17]. Graphical representations of program execution path traversal, in the form of control flow graphs, have been presented as a method for metamorphic malware detection. The proposed method extracts the CFG from a disassembled PE file, which is used to generate an API call graph and converted to a feature vector for processing. The proposed method attained optimum results using a Random Forest classifier to achieve 97% accuracy rate [18].

Android system call sequences have been proposed as an enhanced approach to malware detection and address the shortcomings encountered by traditional methods, which may be evaded as a result of obfuscation techniques. System calls are used to request operating system kernel level services via the application layer or user level processes. System calls may indicate a transfer from user mode to kernel mode to facilitate sensitive operations, such as hardware resource access, device security, network and memory access, inter-process communications, and read and write activities. System calls are an effective mechanism for malware behavior profiling as they are resilient against common evasion practices employed by modifying control flow graphs, opcode sequences, and API calls to obfuscate existing malware. Research conducted on the use of system calls for malware detection demonstrated a 97% detection accuracy in comparison to previous studies [13].

Feature selection techniques are an important step for removing redundant data, enhancing the learning and testing timescales, as well as the overall accuracy of ML for malware detection. Techniques frequently used for feature selection include Chi-square, Fisher Score, Gain Ratio, Information Gain, and Uncertainty Symmetric. The Fisher score technique demonstrated superior results for detecting unknown malware in a study using the Random Forest classifier to detect opcode occurrences [51]. Opcode based features require disassembly first and are therefore not optimized for dynamic ML detection mechanisms. Others propose using sub-signature n-gram term frequency extracted from binaries, using the Information Gain technique and classified with Support Vector Machine, for metamorphic malware detection. The authors claim a 99% detection accuracy rate and an improved performance compared to commercial anti-virus products [30].

A survey of the efficacy of techniques leveraged by machine learning for classification and detection of malware illustrates the efficiencies of employing stochastic modeling techniques for heuristic malware analysis methods and is recommended as an effective method for detecting metamorphic malware variants. Research applying Hidden Markov Models (HMMs) in the training and testing stages demonstrates the effectiveness of malware prediction based on sequence observations, additionally emphasizing the overall enhanced performance of models using system API calls compared with Op-code sequences for detecting malicious activity [46].

Previous studies discuss a HMM-based detector identifying spawned malicious software, which had effectively evaded antivirus software detection; however, additional research demonstrated metamorphic malware successfully evading detection by HMM detectors and commercial antivirus products. ML-based classifiers have similarly proven susceptible to adversarial attacks when purposely generated modifications are introduced into datasets, and researchers propose the inclusion of evasive samples in the training dataset to enhance the overall classification performance [58].

2.2.2 DEEP LEARNING MALWARE DETECTION

Deep learning-based systems have been proposed as an enhancement to traditional ML techniques, for detecting known and unknown malware. Deep learning architectures are deemed as an efficient and scalable method with improved feature engineering capabilities. An experiment using deep learning demonstrated improved results compared to a previous study using manual featuring engineering and the Random Forest method. The study used opcode frequency for feature selection and classification, and using autoencoders and deep neural networks, the outcome was 99.21% accuracy and 0.19% false positive rate. Autoencoders are unsupervised learning algorithms, and the datasets are not dependent on labels, thereby mitigating the manual feature engineering requirement of supervised learning methods [47].

Stacked autoencoders (SAE) have been used for feature extraction from behavior graphs, constructed around deep learning models, to improve malware detection rates compared to earlier studies. The behavior graphs consisted of security critical and related operations of API calls independent of their order. The intention of the graph is to learn potential malicious actions from the combined API calls associated with operating system resources and gain detailed insight into behaviors. Every unique API call behavior is transformed to a binary vector and serves as input to the SAE model. The SAE consists of multiple layers of sparse autoencoders for feature extraction and conversion, to compact and reduce the number of features for optimal representation, and has been demonstrated to improve the accuracy rate for malware detection [60].

Deep learning of call graphs has been proposed as an efficient method for metamorphic malware detection. The approach was inspired by results in computer vision research and the benefits of deep learning image recognition and automatic feature learning. The researchers in Ref. [66] state that obfuscation techniques applied by metamorphic malware alter opcode sequences and byte n-grams; however, functions and calling relationships remain constant after obfuscation, positing that call graphs are more effective for metamorphic malware classification than opcodes and byte values. The proposed method employed a deep convolutional neural network (CNN) for feature learning and classification from malware call graph images, generated from disassembled and PE malware datasets. The reported test accuracy for the PE dataset was 96% and 94.35% for the disassembly dataset. Additionally, CNN's have demonstrated successful results in classifying malicious Android applications using local system calls and their co-occurrence as features. The study by Ref. [33] leveraged NLP techniques to transform system calls, obtained from dynamic analysis, to numerical vectors as input to their learning model. A vocabulary consisting of sequential system calls was used to represent malicious application behavior to produce an accuracy score between 75% and 80%.

Deep learning frameworks have been used to successfully identify unknown malicious Android variants using Linux system kernel calls obtained from dynamic analysis. Linux system calls have been proposed as a more robust method for malware detection as opposed to Android framework API's as they are independent of version variations within the Android operating system and more resilient against evasion techniques used in API substitution. The core component of Android is provided by Linux, and hundreds of system calls support various operating system functions that deliver services to the Android applications when requested. Dynamic analysis performed by researchers using an Android emulator facilitated the execution of malicious Android malware to trace and harvest the system calls requested by the application. The extracted system calls were mapped to integer nodes to construct an overall behavior graph used for classification [24].

Similarly, Linux kernel system calls have been extracted as features in other Android malware detection research using a CNN for automated learning, resulting in a reported accuracy rate of 93.29%. The system calls were correlated with their neighboring system calls to form the basis of a matrix to represent the run time characteristics, system call dependencies, and overall behavior of both benign and malicious software. The matrix was then transformed into feature vectors to represent the collected data as binary values to be used in the machine learning architecture [1].

2.2.3 ADVERSARIAL MACHINE LEARNING

The security arms race between defenders and adversaries encompasses machine learning techniques that have demonstrable vulnerabilities to adversarial attacks. Research has demonstrated adversarial examples of meticulous interference with training, or test input data may sabotage the output accuracy and integrity of machine learning models. Malicious manipulation of training data to modify or introduce data points, aimed at interfering with feature selection and increasing the number of sample misclassifications, is known as a poisoning attack. Evasion attacks are malicious manipulation of test data intended to evade detection by instigating sample misclassification, while maintaining malicious functionality. The challenge to learning algorithms is identifying feature vectors of malicious samples that are indistinguishable from benign samples, thereby increasing the likelihood of evading detection by malware classifiers. The goal of a malware author in the arms race is to evade detection by transforming malicious features to mimic benign samples. The response of the defenders may be reactive or proactive. A reactive response entails adapting behaviors based on learned responses from an adversary; this approach favors an attacker's objectives and does not prevent unknown attacks. Proactive approaches aim to identify weaknesses by simulating an attacker and implementing countermeasures to potential attacks. This approach enhances defense systems in machine learning and improves the detection capabilities of anomalous and novel behavior [10].

Defending against premeditated efforts to bypass classification systems necessitates a threat model designed around the adversary's goals, knowledge, and capabilities. Source-target misclassification is the objective of an adversary in malware evasion, and the goal of the adversary is to cause misclassification of a malicious sample as a benign program. According to the assumed intelligence that an adversary may obtain about a defense model, attacks are typically categorized as either white-box access or black-box access. White-box access implies an attacker is familiar with the model mechanisms and associated constraints. Black-box access suggests an adversary has no knowledge of a defense model and only has limited opportunities to observe results. The success of an adversarial example is dependent on an attackers capability to introduce imperceptible changes to data capable of deceiving a classifier, such as modifying malware to appear benign while maintaining malicious functionality [14].

Deep reinforcement learning has been used to generate metamorphic malware by employing opcode obfuscation, which is capable of evading detection mechanisms with a 67% success rate. A reinforcement learning environment was set up to emulate Markov decision processing to enhance learning through sequentially inserting junk code instructions while maintaining the malicious functionality of the program. Learning agents work together with the environment in a training process whereby the agent modifies the instruction level feature vector by injecting randomly selected opcodes into a malicious code sequence. The generated output acts as input into a discriminator to determine if the feature vectors are benign or malicious. Multiple learning agents may participate in the training process to generate malicious variants with variable levels of obfuscation originating from a single malicious sample. The newly generated samples serve to enhance detection mechanisms to identify novel or metamorphic variants of malicious software and improve malware normalization capabilities [48].

Deep learning discriminative adversarial networks have been proposed as a solution for the detection and classification of obfuscated and non-obfuscated malware by combining feature sets from Android malware in a deep learning design framework. The proposed network negates obfuscation techniques by constructing a multi-view representation of malware based on API calls, raw opcodes, and permissions to learn to detect unusual and potential obfuscation methods used in malware. The proposed implementation employs a CNN along with two discriminator networks designed to ensure that obfuscation does not bias the learned features, one of the networks learns to identify malware while the other network reverses the gradient of obfuscation learning. The design considers malware detection and obfuscation as two opposing tasks, and non-obfuscated training data are augmented with obfuscated data to enhance the learning process. The proposed method demonstrated that combining multiple features improves malware detection rates in comparison to single feature detection rates [36].

2.2.4 GENERATIVE ADVERSARIAL NETWORKS

Deep neural networks have proven vulnerabilities to adversarial attacks demonstrated by generative adversarial examples using obfuscation to successfully evade detection by malware classifiers and cause high misclassification rates in both white-box and black-box scenarios [40]. Additionally, recurrent neural networks (RNN) have been used to generate adversarial examples to evade detection by introducing inconsequential API's to an existing sequence and using a substitute RNN for training in a black-box scenario [25].

Deep reinforcement learning has also been used in black-box attacks to generate evasive malware. The attack was aimed at soliciting classification feedback from a static PE classifier as either benign or malicious and required no knowledge of the model structure or features [2]. The reinforcement model includes an agent which interacts with the detection environment in a contest setting to learn which sequence of PE header metadata changes are likely to evade detection. However, this technique did not perform as well as other approaches to generating adversarial samples and attempts at code obfuscation in a number of cases interfered with functionality.

Reinforcement learning attacks have also been performed against graph neural networks for malware detection, using code obfuscation techniques and control flow graphs as input. Several attack scenarios were explored based on injecting ineffective instructions into binaries without altering the original behavior, in the form of semantic nop instructions [64]. The reinforcement model was deemed suitable for attacks against graph neural networks targeting the node features and graph structure by inserting semantic nops into the control flow graphs in sequence. The methods investigated in the attack scenarios include training the learning agent to repeatedly select code blocks and dead instructions to modify the malicious input, manipulating graphs by inserting dead instructions into randomly selected blocks and using a gradient-descent approach and a hill-climbing approach to evaluate and compare each scenario for efficacy at detection evasion which varied between 45.58% and 100%.

Besides, other initiatives employing generative adversarial network (GAN) algorithms demonstrated successful results for evading black-box learning algorithms using generated malware examples [26]. The method proposes transforming original input samples into complex and flexible adversarial output examples utilizing a generative neural network. It was developed on top of a GAN that relies on a discriminative model to differentiate between generated and genuine samples and a generative model that is trained to deceive the discriminative model to result in misclassification of generated samples as genuine. The architecture contains feed-forward neural networks consisting of a generator and a substitute detector, which work together against a black-box malware detector. The adversarial examples were generated for binary features, given their propensity for accuracy in malware detection, and generation occurs dynamically based on feedback from the black-box detector. The dynamism of this method contrasts with static gradient methods typically used to generate adversarial examples, and it can perform complex transformations along with efficient retraining capabilities.

The generator in MalGAN serves to perform transformations of Windows API feature vectors into diverse adversarial adaptations and introduce random noise to malware by generating feature quantities to cause benign classification [29]. The noise is a randomly selected binary number from the range [0,1] and is concatenated with the malware feature vector to form an input vector. A multi-layer feed-forward neural network receives the input vector and produces an output from the last layer that uses a sigmoid activation function to restrict the range to [0,1]. The objective of the substitute detector in MalGAN is to fit the black-box detector and learn the malicious and benign classification principles that are used for training the generator. Input, in the form of feature vectors, is processed through a multi-layer feed-forward neural network for classification by the substitute detector that uses benign data and adversarial examples from the generator as training data. MalGAN offers a flexible approach for testing the efficacy of malware detection algorithms and enhancing training data for detecting malicious code. However, maximizing the potential of this method necessitates diverse feature quantities for detection learning and limiting samples to a single malware, to facilitate improved performance and realistic attack scenarios.

Moreover, GAN has also been recommended as a potential solution to zero-day malware attacks by generating new samples and signatures evolved from existing malware and expanding on previous research using PE header files for detecting malware families [37]. The classification process improved on earlier research with automated and enhanced feature extraction accuracy provided by long short-term memory (LSTM) and a CNN. Byte code sequences from the File, MS DOS and Optional Header fields provided for features and dependency identification from the PE file. The generator in the proposed method used seven fully connected layers aimed at minimizing the objective function and generated adversarial examples resembling the original samples albeit with minor modifications. The discriminator was trained to maximize the objective function and included four fully connected layers. The generator and discriminator construction presents an alternative approach to the MalGAN implementation. The classifier was trained with genuine and adversarial examples to be able to detect novel malware and demonstrated an improvement on the evaluation metrics of a comparative raw byte method.

Recent research proposed a convolutional GAN framework for generating adversarial examples capable of evading third-party malware detectors by incorporating methods to modify program process execution [67]. The proposed implementation combined additional components with a GAN structure, namely, PE parser for extracting and transforming features into binary vectors, and PE editor to follow the perturbation path from the generator and increase the evasive potential of adversarial examples. Runtime system functions and DLLs were extracted from malware and benign programs to construct a feature set of binary vectors that served as input to the generator along with Gaussian noise. The generator utilized a transposed CNN with four categories of layers to compose a unique representation of malware from an explicit set of features aimed at evading detection. The use of a PE editor introduced an additional layer of complexity via encryption. Output from the generator was used by the PE editor to create an adversarial sample. This process is repeated until a new generated adversarial sample manages to evade the detection. At this juncture, the PE editor creates the optimized adversarial sample. Consequently, the generator was enhanced with the feedback from the discriminator in this process and the evaluation results demonstrate an average decrease of 44% in the detection rate and an increase of 55% in the evasion rate.

Recent research has also attempted to approach the problem through the visualization of malware. GANs have been applied to learn the visualized malware features and to generate novel samples in response to the visualized features to enhance identification of structural correlations of the malware. Byte code sequences from malware binaries were used to form the vector, which is interpreted as grayscale images when processed by the GAN [52]. Additionally, images constructed from malware API call sequence n-grams and related term frequencies have been used in malware GAN research to generate synthetic malware [8]. An improved Wasserstein GAN was utilized to generate synthetic images that were decoded from visual representations to n-gram sequences and term frequencies to enhance behavior-based malware classification.

2.2.5 RELATED WORK

Great effort to seek solutions for metamorphic malware detection is shown in the recent literature [5,62]. In order to improve the positive results of GAN in creating malware mutants, researchers [65] have shown the possibility of generating new malicious mutants using opcode obfuscation created with a deep convolutional GAN (DCGAN), which was initially designed to create images. This included the generation of synthetic features for Android APK files that were able to go undetected by detectors, using a modified DCGAN, together with an algorithm to enable optimum adjustment of the opcode frequency, intended to maintain operational functionality. Novel created APK files were repackaged by employing original features along with other selected features, as well as optimally inserted opcodes, to generate functional packages that were effective in going undetected by four classifiers and VirusTotal. The novel APK files were thereafter employed for training existing detectors to enhance their capability of discovering obfuscated mutants of malicious software.

Furthermore, to tackle malware mutant imbalances in datasets as well as increase overall sample population, and to serve as an utility for data augmentation, generative models have been presented. Reference [15] indicates that the classification accuracy of a malicious mutant is decreased by a small sample representation of the malicious family. An improved classification accuracy of malicious groups was shown by their study with a low representation within a dataset by the conversion of Android opcodes to grayscale images. A DCGAN was employed in creating new samples with results showing an increase in the F1 scores of an independent CNN classifier as against previous scores prior to data augmentation.

A related research [32] intended to enhance diversity of dataset as well as classification performance also used a DCGAN architecture with bytecode malware image representations appearing as RGB images. This employed an 18-layer deep residual architecture based on CNN for classifying the synthetic samples to show a boost in the classification accuracy of unseen data. Furthermore, DCGAN has also been employed with behavior features derived from Android intents in order to improve the defense mechanisms against latent malicious behavioral sequences [27]. These intents comprise of operation descriptions and are employed in initiating send and receive communication services. To generate behavior logs of both cleanware and malware, dynamic analysis was used to serve as input to the DCGAN so as to discover anomalous behavior and malicious activity. The researchers postulate that the fine-tuned pattern analysis and discovery capabilities of DCGAN result in better performance in terms of accuracy in comparison to other machine learning methods.

The research presented in Ref. [27] used only benign samples, and this was done for the fingerprinting of non-malicious behavioral activities and served as the input source for the DCGAN to create new behaviors, which they presuppose may represent likely malicious patterns. The researchers carried out the study using dynamic analysis; they, however, left out the image creation process in their discussion. Reference [32] used an existing malware image database for their research and conducted classification using a deep residual network; however, deep classification networks are designed for large-scale image recognition and require specific image dimensions. Reference [15] used image data augmentation methodologies and double-layer CNN based on the generative model and did not implement deep models for classification. The features used in this study represent the runtime system calls from verifiable metamorphic malware to enhance detection rates of novel mutations. All researchers employed the F1-score during evaluation. We present a summary of the

Table 2.1
Related Work Summary Illustrating Features, Datasets, and Classification Algorithms and Evaluation Metrics Employed

Source	Features	Dataset	Classification	Evaluation Metric
[15]	Opcodes	20,000 data in AMD 4,000 Debrin	CNN	Accuracy, precision, recall and F1-score
[27]	Intents	Benign samples	CNN, DCGAN, LSTM	Accuracy, precision, false positive rate and F1-score
[32]	Bytecode	2,949 Malicious samples	DRN	Accuracy, precision, recall and F1-score
[65]	Opcodes	10,021 Benign, 10,035 Malicious samples	CNN , KNN, SVM and Kaggle-RF	False negative rate, precision, recall and F1-score

related work in Table 2.1. The features used in the study discussed in this chapter represent the runtime system calls from verifiable metamorphic malware to enhance detection rates of novel mutations.

2.3 METHODOLOGY

This study proposes to use a DCGAN to generate novel color images based on behavioral features dynamically extracted from a unique metamorphic malware dataset. The primary aim of the experiments is to generate novel metamorphic Android malware features aimed at improving the identification and classification of potential novel malicious variants. The process of malware feature generation is outlined in this section and shown in Figure 2.1. It follows the life cycle of the evidence-based experiments designed in accordance with the scientific method described in Ref. [57].

2.3.1 DATASET

The dataset used in this research consists of both malicious [4] and benign samples consisting of games and utility applications obtained from the F-Droid online repository [31]. The balance between the benign and malicious samples contributes to the development of an unbiased classifier. The malicious class is comprised of novel mutant samples generated from three different Android malware variants, namely, Dougalek,[1] DroidKungFu[2] and GGTracker[3].

2.3.2 DYNAMIC ANALYSIS

User interaction with the malicious software was simulated with the MonkeyRunner[4] utility available in the Android Software Development Kit (SDK) to trigger random

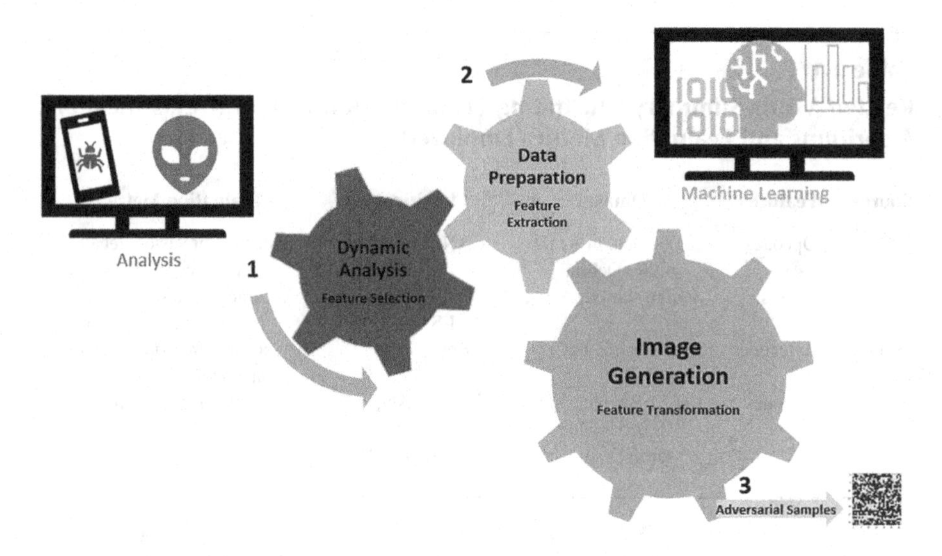

Figure 2.1 Metamorphic GAN system architecture.

events, such as keystrokes and screen activity, to initiate interaction with the malicious software and enable execution [53]. The data required for the experiment in this research were acquired from the system call traces of all the dataset Android package kit (APK) samples that were installed on an Android virtual device (AVD) within a Docker container hosted on the analysis VM using the Linux diagnostic utility Strace.[5]

2.3.3 DATA PREPARATION

The system call trace files were sanitized to remove all data except for the system calls in the order of occurrence, rendering a list of thousands of chronological system calls for each sample. The behavioral characteristic for each sample is contained in the Strace log file, which is subsequently transformed into a n-gram feature representation using term frequency and inverse document frequency (TFIDF) vectorization. TFIDF is a numerical model frequently used to measure the statistical importance of a word or sequence of words within a document. Word count occurrences are measured by term frequency within a document and are represented as a n-gram, which is divided by the sum of unique n-gram term frequencies as a process of normalization. Inverse document frequency (IDF) scales up rare n-gram occurrences within a document and downscales n-grams that have a high occurrence rate and therefore a lower significance [54]. The Strace log files represent a document, and the system calls represent n-gram terms for the purpose of the experiments conducted.

2.3.4 IMAGE GENERATION

Visualizing malware execution patterns as feature images of malicious behavior has demonstrated high accuracy rates for detection and classification using deep learning technology. This technique involves mapping malware features to pixel intensities, which represents the malware behavior as an image. An image is generated from a feature matrix and mapped to image color channels to create a fingerprint representing malware behavior [55]. The images created for the purpose of the experiment are generated from the TFIDF 2-gram matrix to construct the 32×32 height and width color images. An illustration of the conversion of system call n-grams to images is shown in Figure 2.2.

2.3.5 ADVERSARIAL SAMPLES

The images created from the malware sample features in the dataset were used to generate novel samples utilizing the DCGAN framework to generate new images for each malicious family. DCGAN is an architecture class of the original GAN framework that uses convolutional layers in the generator and discriminator models

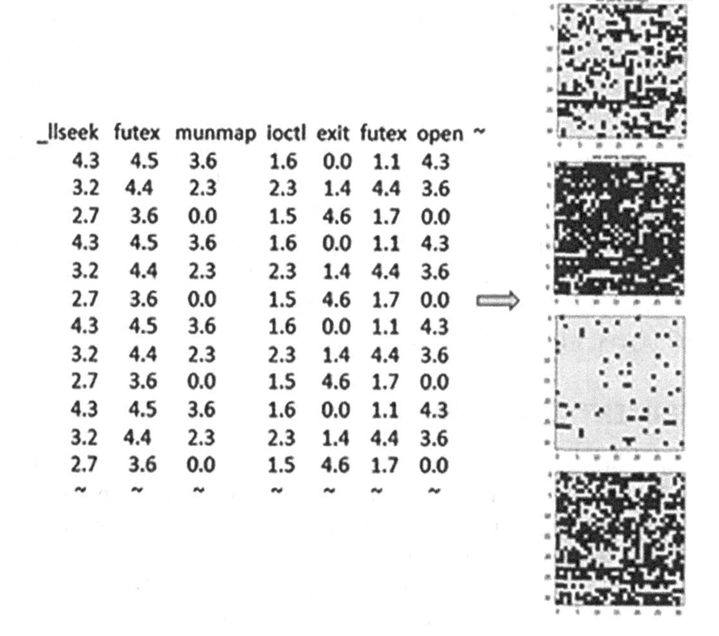

Figure 2.2 n-gram features to image transformation.

to learn reusable feature representations to produce novel output. The purpose of the generator is to create fake images using a latent vector along with an original image as input. The discriminator serves to classify an input image as real or fake in a competitive configuration where both the models attempt to improve based on their respective performance feedback [42].

2.3.6 CONVOLUTIONAL NEURAL NETWORK (CNN)

CNNs have demonstrated high accuracy and performance rates compared to other models that employ feature extraction classification techniques. CNNs were initially designed for computer vision tasks and take images as an input of *n*-dimensional shape corresponding to the length, width, and channel of an image, and uniformity in dataset sizes is required. The structure of a CNN includes an input layer, several subsampling layers, and fully connected layers, which convert the *n*-dimension features to satisfy the classification criteria [16].

CNNs are a biologically inspired architecture designed for classifying images into predefined classes, and the original design is based on the layers and cells of the visual cortex of the brain. The design comprises convolutional and pooling layers that are stacked on top of each other. Images serve as input to a feedforward neural network that processes information from input to output in one direction only being processed through several fully connected layers. Image feature extraction and learning occurs at the convolutional layers where feature maps are formed from neurons that are connected to neighboring neurons in the preceding layer through learned weights.

A new feature map, calculated from the convolved input and learned weights, is then processed through an activation function. The pooling layers serve to reduce input distortions and are configurable as average or max pooling for transmitting the input values to the next layer. Max pooling selects the largest value in a field while average pooling calculates an average of all the input values. High-level reasoning is performed to understand the feature representations extracted by the pooling and convolutional layers, which collectively form fully connected layers [44].

2.4 EXPERIMENTAL DESIGN

The virtual machines (VMs) implemented for the purpose of the experiment were hosted with Microsoft Hyper-V Manager in accordance with objectives of the experiments. Hyper-V technology provides isolated virtual partitions with each VM running dedicated hypervisor resources and own processes with restricted networking functionality to protect the host system and prevent inter VM communication [20]. Virtualization technology provided a dedicated protected environment that enabled interaction with malicious software within the boundaries of legal and ethical responsibilities and contain any potential malicious activities. VMs were created to perform the following functions:

Table 2.2
Hardware and Software Specifications of the Experimental Environment

Role	OS	CPU	RAM	Networking
Host	64 bit Windows 10 Professional	Intel Core i7	16.0 GB	Enabled
Analysis VM	Ubuntu 18.04 LTS	Intel Core i7	Dynamic Allocation	Disabled
Machine learning VM	Ubuntu 18.04 LTS	Intel Core i7	Dynamic Allocation	On demand

- **Analysis VM**: This is the machine used for analysis that hosts the Android emulation software that collects malicious and benign artifacts by performing dynamic analysis. It has no provision for internet connectivity and is disconnected from the virtual switch.
- **Machine learning VM**: This hosts the necessary software to perform data preparation and machine learning model evaluation, internet connectivity provided as and when required.

A summary of the environment is outlined in Table 2.2.

2.4.1 EXPERIMENTAL SETUP

The experiments are structured around a binary classification problem with two classes, namely, benign and malicious, for which the classification model delivers discrete or continuous outputs. The discrete output predicts the class or label of an unseen sample from the test set, and the continuous output estimates the probability of the sample belonging to one class or the other. In order to assess the model's prediction ability to identify novel malicious samples, the dataset is split using stratified *k*-fold cross-validation to ensure the subset proportions are reflective of the training set proportions. This is to ensure that predictive features are selected from the training set as opposed to the complete dataset and are intended to mitigate against predictive bias and overfitting. For the purpose of the experiments, 10-fold cross-validation was employed as recommended for real-world datasets and unbiased error prediction [7]. The dataset is structured into training, validation, and testing sets comprising the images derived from the dataset and novel variants generated by the DCGAN. The performance measurement metrics employed include accuracy, precision, recall, and F-measure/F1-score [56].

Two experiments were carried out in this study to answer the second research question, "How effective are generative neural networks for enhancing the detection of metamorphic malware using behavioral characteristic?". The first experiment was aimed at assessing the evasive effectiveness of the generated samples using a CNN classifier. The following steps were performed during the experiment:

1. Load the data into the CNN
2. Data preparation and augmentation
3. Build the classification model with the training set
4. Perform validation on the validation set
5. Measure the training and validation accuracy
6. Perform binary classification to predict the class using unlabeled images for each malicious family test set
7. Perform binary classification to predict the class using the unlabeled benign samples test set
8. Calculate the classification metrics

The second experiment was aimed at assessing the effectiveness of the CNN classifier with a new dataset that includes the test set from the first experiment in the training and validation sets and a newly generated batch of malicious samples for each family produced by the DCGAN. The following steps were performed during the experiment:

1. Load the data into the CNN
2. Data preparation and augmentation
3. Perform training on the training set
4. Perform validation on the validation set
5. Measure the training and validation accuracy
6. Perform binary classification to predict the class using unlabeled images for each new malicious family test set
7. Calculate the classification metrics

2.4.2 BEHAVIOR FEATURE EXTRACTION

Dynamic analysis of the malicious APK dataset samples was achieved with a docker container of the integrated Android malware analysis framework utility[6], and specifically the modified implementation of DroidBox[7] included in the framework. The analysis VM hosted the latest Docker engine and Andropytool container along with Android SDK version 4.1.3. Droidbox is an Android application dynamic analysis tool, and the customized version included with the docker image includes Strace, within the Android emulator for system call capturing in addition to a higher number of user actions simulated by MonkeyRunner for a duration of 300 seconds. Droidbox hosts an armeabi-v7a emulator architecture running Android 4.1.1 for AVD devices and enables real-time detailed analysis of application behavior. Strace facilitates monitoring and logging of all application system calls at the kernel level while running from the zygote process, including launched services, circumvented permissions, loaded DEX classes, and file read and write activity. The log file generated by Strace includes all timestamped execution events for the dataset APK's received as input by the tool [34].

The data samples were provided as input to the Andropytool docker container with a limited argument to only launch Droidbox within the framework to conduct dynamic analysis. The process includes executing Droidbox that launches an emulator and installs each APK individually for a duration of 300 seconds of user-simulated activity. Application behavioral activity is captured and output to CSV, JSON, and text format log files. Once the capture is complete, the AVD device is reverted, subsequent applications are installed, and the process is repeated. The pertinent log files for the purpose of the experiment were the Strace logs, which include all system calls with associated arguments and process identification number (PID). The log files were processed to remove the arguments, PIDs, timestamps, and blank lines between system calls, to render a list of system calls in sequential order reflecting the behavior of each benign and malicious sample from the dataset.

The average runtime process is 6 minutes for each sample in the dataset, including the emulator revert time and user activity stimulation process. The average number of system calls for the entire dataset is 38,544 per sample. The Strace logs were stored to disk under their variant name directory as per the dataset structure.

2.4.3 WORDS TO IMAGES

Each refined log file was processed as a document of words signifying the system call sequence for every APK file in the dataset. A vocabulary of 2-grams is constructed from the documents using the machine learning Python module Scikit learn [38]. Feature extraction of n-grams from the documents was achieved with the TFIDF Vectorizer technique. The maximum features were capped at $1,024$ to satisfy the desired image dimensions ($32 \times 32 = 1,024$), and sublinear term frequency was used to counteract the significance of highly repetitive system calls or deliberate obfuscation attempts at inserting redundant calls [9].

A modified version of a Python malware feature extraction script to include the aforementioned vectorizer parameters was executed in each sample set directory to produce a 2-dimensional data structure of the system calls and associated frequency of each sample [12]. The output for each malware family and benign set were saved to disk in a csv matrix generated by the script. The rows of each matrix are a 1-dimensional behavioral vector representation of the data samples and are converted into a 2-dimensional image representation, where each TFIDF value is transformed into an 8-bit RGB pixel value ranging between 0 and 255. As image uniformity in shape is required by CNNs, all the experimental images were created in the same size in preparation for being processed.

2.4.4 SYNTHETIC IMAGES

The image representations of the malware dataset are used as input to the DCGAN in order to generate novel samples required for the experiment. The input images for the experiments are 3-by-32-by-32 where 3 denotes the number of channels for RGB images. The objective function defined in Eq. (1.1) is used to optimize the DCGAN.

$$\min_{G}\max_{D} V(D,G) = \mathbb{E}_x \sim p_{\text{data}}(x)[logD(x)] + \mathbb{E}_z \sim p_z(z)[\log(1-D(G(z)))] \quad (2.1)$$

The generator (G) and discriminator (D) compete in a minimax contest with the value function $V(D,G)$. $\mathbb{E}_x{\sim}p_{\text{data}}(x)$ signifies genuine data distribution, whereas $\mathbb{E}_z{\sim}p_z(z)$ denotes fake data distribution z. The objective of the D is to maximize the probability of correct classification between genuine and fake data, denoted by $\log D(x)$, while G's objective is to minimize the probability of accurate classification by D, indicated by $[\log(1-D(G(z)))]$ [22]. The discriminator used in the DCGAN consists of strided layers that apply 2-dimensional mathematical operations over an input signal of several input planes to output a scalar probability of the input $3 \times 32 \times 32$ for the experiment images. The generator in the DCGAN includes several layers that apply 2-dimensional transposed mathematical operations on an input latent vector to output a transformed $3 \times 32 \times 32$ RGB image.

The discriminator is essentially a down sampling CNN, whereas the generator is an up sampling CNN, and both make use of batch normalization functions that enable the use of higher learning rates [43]. A Pytorch DCGAN[8] was utilized to generate the images for the experiment. A modified version of the generator was employed to function with 32×32 images. The malware dataset images served as input to the DCGAN in separate cycles per variant to produce synthetic images for each family. The objective of the generator is to convert latent space to images with the same size as the input images. This is accomplished through the process of the transposed convolutional layers, batch normalization, and activation functions, which is then output through a tanh function to revert the image to the input data range. The discriminator serves as a binary classification network that processes the images through several convolutional layers, batch normalization, and activation functions to determine a probability of the authenticity of the input image as real or fake.

The DCGAN framework was initialized to use the custom dataset folders and produce individual synthetic image outputs at the completion of each epoch. The training is performed in two stages. Stage one is designed to maximize the probability of accurate classification by D, denoted as $\log(D(x)) + \log(1-D(G(z)))$. Training batches of genuine samples complete a forward pass through D to calculate the loss denoted as $\log(D(x))$ and a backward propagate the gradient of the loss function with respect to the neural network's weights. Then, the losses and gradient accumulations for a batch of synthetic samples, produced by G, are calculated by completing a forward pass through D, denoted as $\log(1-D(G(z)))$ and a backward pass, respectively. Stage two is intended to maximize $\log(D(G(z)))$ by optimizing G's steps based on G's stage one losses and backward propagate the gradient of the loss function with respect to the neural network's weights.

Table 2.3
DCGAN Hyperparameters

Hyperparameters	Values
Number of iterations	20
Learning rate	0.0002
Beta1	0.5
Random seed	Yes
nz input vector	100
Batch size	1

DCGAN uses the binary cross entropy loss (BCELoss) function to enhance learning for D and G with two separate Adam optimizers for each network. The BCELoss function used in the PyTorch DCGAN calculates the log components required for both D and G and is defined in (2.2).

$$\ell(x,y) = L = [l_1,\ldots,l_N]^T, l_n = -[y_n.\log x_n + (1-y_n).\log(1-x_n)] \tag{2.2}$$

The component of the equation to calculate is specified by y during the training, which is mutable and defines the labels. Training is complete when G has exhausted all possibilities of generating new samples or when D can no longer differentiate between genuine distributions and synthetic outputs from G. The average runtime for the selected iterations was 6 minutes per malicious dataset variant. The hyperparameters configured for the DCGAN are consistent with the original research by Ref. [43], with the exception of the number of iterations and batch size as illustrated in Table 2.3.

Sixty synthetic image samples were generated with the DCGAN from source images representing each malicious variant equally, and a sample is illustrated in Figure 2.3.

2.4.5 IMAGE CLASSIFICATION

A CNN-based image classifier is employed to assess the evasive potential of the newly generated images produced by the DCGAN. The CNN is trained using the original and synthetic images. It aims for binary classification by estimating the probability of the generated test images being malicious or benign. Alternatively, multi-class classification can be sought to further classify several different classes [49].

The implementation involves providing the images to the input layer which are formulated for feature extraction by the convolutional layer through resizing. The convolutional layer filters the images to discover and calculate features to perform

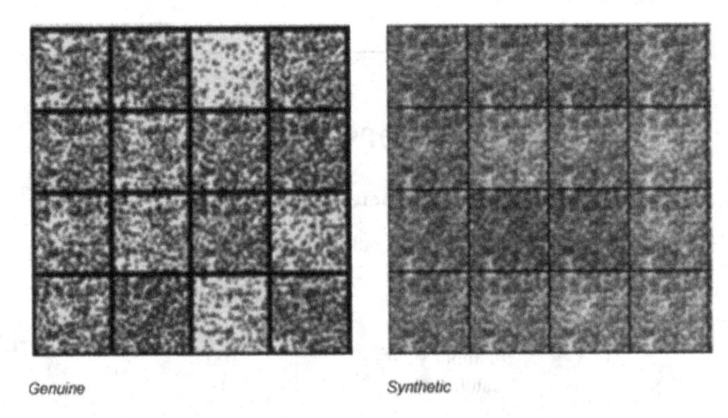

Figure 2.3 Sample image from 60 synthetic image samples generated with the DCGAN from source images representing each malicious variant equally.

feature mapping during testing. The extracted features are transferred to the pooling layer that shrinks the image while maintaining the maximum relevant information through max pooling. This information is passed to the activation layer that mathematically stabilizes the learned values and passes them to the fully connected layer that translates and categorizes the high-level filtered images to classes [50].

Comparable with the research by Ref. [15], this experiment is aimed at the generative capabilities of a DCGAN using malware features as opposed to image classification architectures, and therefore, the model implemented resembles a simple convolutional network model. A Keras CNN image classification network was implemented for the purpose of the experiment. Data augmentation was applied to safeguard against the model processing the same image twice and is intended to optimize the use of the small dataset. Random image transformations counteract overfitting and improve model generalization, and the dropout inhibits the exact patterns being viewed by a layer twice. The stack consists of three convolutional layers with a ReLU activation function and max-pooling layers with two fully connected layers. Feature maps are converted from 3-dimensional to 1-dimensional feature vectors by the flatten function, and sigmoid and binary cross-entropy are implemented for binary classification.

The original sample dataset containing the images is structured into two classes representing benign and malicious samples using 10-fold cross-validation to split the dataset into train, validation, and test folders. A modified version of a Keras Jupyter Notebook was implemented for the experiment and configured to load the data from local folders within the ML VM to generate labels from batches of the images. The data were prepared utilizing data augmentation and model generators to load and process the data in batches from the subfolders and train the model. The validation data are used to measure the training performance accuracy, and the test data contain images previously unseen by the model which are used to predict the probability of class membership, either malicious or benign. The CNN model hyper-parameters are listed in Table 2.4.

Table 2.4
CNN Hyperparameters

Hyperparameters	Values
Image size	32
Epoch	20
Batch size	10
Test size	20
Input shape	32,32,3

The experiments were then conducted, and the results are analyzed in the subsections below to answer our research questions.

2.5 RESULTS AND DISCUSSION

This section discusses the detail of the experiments, and the results obtained to evaluate the efficacy of a generative network aimed at producing novel samples using metamorphic malware features. DCGAN generator and discriminator networks learn to improve based on updated learning parameters provided by the gradient descend algorithm. The input to the discriminator is an image, and the training objective is to increase the probability of accurate classification as authentic or an imitation, thereby maximizing $\log(D(x)) + \log(1 - D(G(z)))$, where x represents the image, and z represents the generator function mapped to the latent vector. In this case, $D(x)$ performs as a conventional binary classifier. The generator is trained to minimize $\log(1 - D(G(z)))$ aimed at producing improved imitation samples, and the loss is calculated as $\log(D(G(z)))$. The discriminator loss is determined by the sum total of all sample losses by classifying the generator output [41].

2.5.1 ASSESSING THE EVASIVE EFFECTIVENESS OF THE GENERATED SAMPLES USING A CNN CLASSIFIER

The first experimental cycle (experiment 1) targeting the original dataset without any synthetic images provides a baseline to measure the performance of the CNN. The model correctly predicted all the benign samples with a 99.7% probability prediction average across the samples. A single malicious sample was misclassified as benign, while the remaining samples were correctly predicted as malicious with a 99.9% probability prediction average. Subsequent cycles involved replacing the original malware in the test folder with synthetic samples from each family to conduct individual classification rounds for each variant. The combined results for the synthetic samples demonstrate a reduction in the overall prediction probability averages for the synthetic malicious samples with 51% of the samples misclassified as benign. This experiment demonstrated that synthetic samples produced by a deep generative network are successful at evading detection.

2.5.2 ASSESSING THE EFFECTIVENESS OF THE CNN CLASSIFIER WITH A NOVEL DATASET INCLUDING A NEWLY GENERATED BATCH OF MALICIOUS SAMPLES FOR EACH FAMILY PRODUCED BY THE DCGAN

This stage was aimed at evaluating the efficacy of the CNN to detect novel samples by creating a new dataset to include the test set from the first experiment in the training and validation sets and a newly generated batch of malicious samples for each family produced by the DCGAN for the test set. The combined results for this experiment (experiment 2) demonstrate a 91% accuracy detection rate of the newly generated malicious samples and an overall probability prediction average of 95%, demonstrating an improvement on the results from experiment one. An interesting anomaly was observed between the first experiment and the second experiment for the DroidKungfu variant with over 50% misclassification in experiment one to 100% correct classification in experiment two. This result was consistent in all the tests conducted in experiment 2.

The results (experiment 2) indicate an overall improvement in comparison to experiment 1, in particular the F measure, indicating the balance between recall and precision performance with an increase of 0.31% toward the optimal value of experiment one as listed in Table 2.5. Experiment 2 demonstrated that the inclusion of synthetic samples in training datasets improves detection rates of novel malicious samples. The experiment was executed for ten repeated cycles per family capturing the highest evasive results, and the average timescale for each testing and validation cycle was 30 seconds with an additional average of 5 seconds for each testing cycle when conducted on the ML VM allocated with four virtual processors. The results obtained from the experiments were used to evaluate the efficacy of a DCGAN for producing novel synthetic malware behavior features and ultimately the effectiveness for enhancing detection capabilities.

Table 2.5

Comparison of Experiment 1 and Experiment 2 Highlighting the Improved Performance Utilizing Synthetic Samples in the Dataset

Performance Metrics	Experiment 1	Experiment 2
Accuracy	0.73	0.95
Precision	0.97	0.98
Recall	0.48	0.92
F1-scores	0.64	0.95

2.5.3 EVALUATION

The overall objectives of this research were to identify the essential features of metamorphic malware that may improve detection capabilities and evaluate the efficacy of deep generative neural networks for enhancing the detection of metamorphic malware using the identified behavioral characteristics. To achieve these objectives, a literature review was conducted to compile background information from a range of sources on related research into metamorphic malware and machine learning techniques for malware detection. Additionally, the background information gathering served to identify appropriate tools and techniques for structuring the design of the experiments performed in this research project. The literature review identified Linux kernel level system calls as a reliable method for malware detection due to their independence of Android operating system version variations and resilience against common evasion techniques. Deep convolutional generative adversarial networks were also identified as an effective method of generating synthetic novel samples due to the fine-tuned pattern analysis and identification capabilities compared to other models. The decision to utilize a DCGAN for this project resulted in the requirement to convert the metamorphic malware features into images as input to the framework and ultimately influenced the classification method for the experiments which is consistent with methods utilized in other related work on this topic.

The research conducted by Ref. [32] demonstrated an overall 6% improvement in F1 score using a DCGAN to augment baseline samples and a deep residual network for classification. A deep classification network was evaluated during this project; however, it was not deemed appropriate due the poor results during training and evaluation due to low population of the research dataset. The dataset used in Ref. [32] was approximately three times larger than the dataset used in this project, and the generative samples were produced over thousands of epochs.

The research carried out by Ref. [65] was aimed at the evasiveness of generative produced samples measured with several different classifier models and did not evaluate the results of retraining classification models including novel samples. The dataset used in the study contained thousands of samples commonly used in research.

The study conducted by Ref. [15] demonstrated an improved F1 score between 5% and 20% across samples using augmented generative samples to measure the effects of using novel variants to improve detection rates using a simple CNN network for classification results. The dataset used in the study contained thousands of samples from two commonly researched datasets. All the aforementioned studies focus on static analysis features that are more susceptible to obfuscation attacks.

The dataset used in this research project was significantly smaller than related studies. However, it is a true representation of metamorphic malware, which was the focus of this work. The training and evaluation accuracy results of the classifier suggest that overfitting may compromise the prediction reliability, despite efforts to mitigate against overfitting with data augmentation techniques and model parameters.

The improved F1 scores produced in experiment 2 demonstrate the potential classification performance improvement by augmenting the data with novel samples derived from the DCGAN and metamorphic malware feature samples.

2.6 CONCLUSION

The experimental results of this project demonstrate the efficacy of machine learning techniques for enhancing the detection capabilities of anomalous patterns. Metamorphic malware is stealthy by design and benefits from several obfuscation techniques to evade detection. The experiment results correlate with related research utilizing generative machine learning to enhance malware detection. However, there were no studies discovered during the background information gathering, which focused on verifiable metamorphic malware features and generative networks.

In addressing the research objective that was to implement and evaluate a generative neural network for enhancing metamorphic malware detection based on behavior profiling, two research questions were posed. The first question was addressed in the related work section and its focus was narrowed down to address metamorphic malware due to the evasive obfuscation techniques and was constricted further for Android applications. The background research demonstrated successful results employing techniques utilizing low-level features that were ultimately used in this project. The second question was addressed in the experimental results, demonstrating the efficacy of a DCGAN using low-level features in improving detection capabilities, which is validated by the improvement in the F1 score between the two experiments conducted.

Future work would benefit from utilizing a larger metamorphic malware dataset to address the high training and validation accuracy results and the concern with overfitting listed in the Results and Discussion section. The experiment could be expanded to incorporate processes to attempt losslessly decoding the generated images back to system call sequences and extended further by including multi-class classification with a larger dataset and evaluating the effectiveness of identifying individual malware families derived from a DCGAN.

Notes

1 Dougalek - https://www.trendmicro.com/vinfo/us/threat-encyclopedia/malware/androidosdougalek.a
2 Droidkungfu - https://www.f-secure.com/v-descs/trojan˙android˙droidkungfu˙c.shtml
3 GGtracker - https://www.f-secure.com/v-descs/trojan˙android˙ggtracker.shtml
4 Monkeyrunner - https://developer.android.com/studio/test/monkey
5 Strace - https://linux.die.net/man/1/strace
6 Andropytool - https://securityonline.info/andropytool-automated-extraction-of-static-and-dynamic-features-from-android-applications/
7 Droidbox - https://www.honeynet.org/taxonomy/term/191
8 Pytorch DCGAN - https://www.pyimagesearch.com/2021/10/25/training-a-dcgan-in-pytorch/

REFERENCES

1. Abada Abderrahmane, Guettaf Adnane, Challal Yacine, and Garri Khireddine. Android malware detection based on system calls analysis and cnn classification. In *2019 IEEE Wireless Communications and Networking Conference Workshop (WCNCW)*, pp. 1–6, Marrakech, 2019.
2. Hyrum S. Anderson, Anant Kharkar, Bobby Filar, David Evans, and Phil Roth. Learning to evade static PE machine learning malware models via reinforcement learning. *CoRR*, abs/1801.08917, 2018.
3. Giovanni Apruzzese, Michele Colajanni, Luca Ferretti, Alessandro Guido, and Mirco Marchetti. On the effectiveness of machine and deep learning for cyber security. In *2018 10th International Conference on Cyber Conflict (CyCon)*, pp. 371–390, Tallinn, 2018.
4. Kehinde O. Babaagba, Zhiyuan Tan, and Emma Hart. Nowhere metamorphic malware can hide: A biological evolution inspired detection scheme. In Guojun Wang, Md Zakirul Alam Bhuiyan, Sabrina De Capitani di Vimercati, and Yizhi Ren, editors, *Dependability in Sensor, Cloud, and Big Data Systems and Applications*, pp. 369–382. Springer, Singapore, 2019.
5. Kehinde O. Babaagba, Zhiyuan Tan, and Emma Hart. Improving classification of metamorphic malware by augmenting training data with a diverse set of evolved mutant samples. In *2020 IEEE Congress on Evolutionary Computation (CEC)*, pp. 1–7, Glasgow, 2020.
6. Babak Bashari Rad, Maslin Masrom, and Suhaimi Ibrahim. Camouflage in malware: From encryption to metamorphism. *International Journal of Computer Science And Network Security (IJCSNS)*, 12:74–83, 2012.
7. Daniel Berrar. Cross-validation. In Shoba Ranganathan, Michael Gribskov, Kenta Nakai, and Christian Schönbach, editors, *Encyclopedia of Bioinformatics and Computational Biology*, pp. 542–545. Academic Press, Oxford, 2019.
8. Vineeth S. Bhaskara and Debanjan Bhattacharyya. Emulating malware authors for proactive protection using gans over a distributed image visualization of dynamic file behavior, 2018.
9. Vineeth S. Bhaskara and Debanjan Bhattacharyya. Emulating malware authors for proactive protection using gans over a distributed image visualization of the dynamic file behavior. *ArXiv*, abs/1807.07525, 2018.
10. Battista Biggio and Fabio Roli. Wild patterns: Ten years after the rise of adversarial machine learning. *Pattern Recognition*, 84:317–331, 2018.
11. Anna L. Buczak and Erhan Guven. A survey of data mining and machine learning methods for cyber security intrusion detection. *IEEE Communications Surveys Tutorials*, 18(2):1153–1176, 2016.
12. Ricardo A. Calix, Sumendra B. Singh, Tingyu Chen, Dingkai Zhang, and Michael Tu. Cyber security tool kit (cybersectk): A python library for machine learning and cyber security. *Information*, 11(2):100, 2020.
13. Gerardo Canfora, Eric Medvet, Francesco Mercaldo, and Corrado Aaron Visaggio. Detecting android malware using sequences of system calls. In *Proceedings of the 3rd International Workshop on Software Development Lifecycle for Mobile, DeMobile 2015*, pp. 13–20, New York, NY, 2015. Association for Computing Machinery.
14. Nicholas Carlini, Anish Athalye, Nicolas Papernot, Wieland Brendel, Jonas Rauber, Dimitris Tsipras, Ian J. Goodfellow, Aleksander Madry, and Alexey Kurakin. On evaluating adversarial robustness. *CoRR*, abs/1902.06705, 2019.

15. Yi-Ming Chen, Chun-Hsien Yang, and Guo-Chung Chen. Using generative adversarial networks for data augmentation in android malware detection. In *2021 IEEE Conference on Dependable and Secure Computing (DSC)*, pp. 1–8, Aizuwakamatsu, 2021.
16. Zhihua Cui, Fei Xue, Xingjuan Cai, Yang Cao, Gai-ge Wang, and Jinjun Chen. Detection of malicious code variants based on deep learning. *IEEE Transactions on Industrial Informatics*, 14(7):3187–3196, 2018.
17. Ammar Elhadi, Mohd Maarof, and Bazara Barry. Improving the detection of malware behaviour using simplified data dependent API call graph. *International Journal of Security and Its Applications*, 7:29–42, 2013.
18. Mojtaba Eskandari and Sattar Hashemi. Metamorphic malware detection using control flow graph mining, 2011.
19. Zhiyang Fang, Junfeng Wang, Jiaxuan Geng, and Xuan Kan. Feature selection for malware detection based on reinforcement learning. *IEEE Access*, 7:176177–176187, 2019.
20. Lee Garber. The challenges of securing the virtualized environment. *Computer*, 45(01):17–20, 2012.
21. R. Geetha and T. Thilagam. A review on the effectiveness of machine learning and deep learning algorithms for cyber security. *Archives of Computational Methods in Engineering*, 28(4):2861–2879, 2021.
22. Ian J. Goodfellow, Jean Pouget-Abadie, Mehdi Mirza, Bing Xu, David Warde-Farley, Sherjil Ozair, Aaron Courville, and Yoshua Bengio. Generative adversarial networks, 2014.
23. William Hardy, Lingwei Chen, Shifu Hou, Yanfang Ye, and Li Xin. DL4MD: A deep learning framework for intelligent malware detection. In *The IEEE International Conference on Data Mining (ICDM)* . CSREA Press, 2016.
24. Shifu Hou, Aaron Saas, Lifei Chen, and Yanfang Ye. Deep4MalDroid: A deep learning framework for android malware detection based on linux kernel system call graphs. In *2016 IEEE/WIC/ACM International Conference on Intelligence Workshops (WIW)*, pp. 104–111, Omaha, NE, 2016.
25. Weiwei Hu and Ying Tan. Black-box attacks against RNN based malware detection algorithms. *CoRR*, abs/1705.08131, 2017.
26. Weiwei Hu and Ying Tan. Generating adversarial malware examples for black-box attacks based on GAN. *CoRR*, abs/1702.05983, 2017.
27. Salman Jan, Shahrulniza Musa, Toqeer Syed, and Ali Alzahrani. Deep convolutional generative adversarial networks for in-tent-based dynamic behavior capture. *International Journal of Engineering and Technology*, 7:101–103, 2018.
28. Ilker Kara. A basic malware analysis method. *Computer Fraud & Security*, 2019(6):11–19, 2019.
29. Masataka Kawai, Kaoru Ota, and Mianxing Dong. Improved malgan: Avoiding malware detector by leaning cleanware features. In *2019 International Conference on Artificial Intelligence in Information and Communication (ICAIIC)*, pp. 040–045, Okinawa, 2019.
30. Ban Mohammed Khammas, Alireza Monemi, Izhairi Ismail, Sulaiman Mohd Nor, and Muhammad Nadzir Marsono. Metamorphic malware detection based on support vector machine classification of malware sub-signatures. 14:1157–1165, 2016.
31. Pei Liu, Li Li, Yanjie Zhao, Xiaoyu Sun, and John Grundy. Androzooopen: Collecting large-scale open source android apps for the research community. In Georgios Gousios and Sarah Nadi, editors, *Proceedings - 2020 IEEE/ACM 17th International Conference on Mining Software Repositories, MSR 2020*, pp. 548–552, United States of America, 2020. Association for Computing Machinery (ACM).

32. Yan Lu and Jiang Li. Generative adversarial network for improving deep learning based malware classification. In *2019 Winter Simulation Conference (WSC)*, pp. 584–593, National Harbor, MD, 2019.
33. Fabio Martinelli, Fiammetta Marulli, and Francesco Mercaldo. Evaluating convolutional neural network for effective mobile malware detection. *Procedia Computer Science*, 112:2372–2381, 2017. *Knowledge-Based and Intelligent Information & Engineering Systems: Proceedings of the 21st International Conference, KES-20176-8*, Marseille, September 2017.
34. Alejandro Martín, Raúl Lara-Cabrera, and David Camacho. Android malware detection through hybrid features fusion and ensemble classifiers: The andropytool framework and the omnidroid dataset. *Information Fusion*, 52:128–142, 2019.
35. Hoda El Merabet and Abderrahmane Hajraoui. A survey of malware detection techniques based on machine learning. *International Journal of Advanced Computer Science and Applications*, 10(1):366–373, 2019.
36. Stuart Millar, Niall McLaughlin, Jesus Martinez del Rincon, Paul Miller, and Ziming Zhao. *DANdroid: A Multi-View Discriminative Adversarial Network for Obfuscated Android Malware Detection*, pp. 353–364. Association for Computing Machinery, New York, NY, 2020.
37. Zahra Moti, Sattar Hashemi, and Amir Namavar. Discovering future malware variants by generating new malware samples using generative adversarial network. In *2019 9th International Conference on Computer and Knowledge Engineering (ICCKE)*, pp. 319–324, Mashhad, 2019.
38. Fabio Nelli. *Machine Learning with Scikit-learn*, pp. 313–347. Apress, Berkeley, CA, 2018.
39. Philip O'Kane, Sakir Sezer, and Kieran McLaughlin. Obfuscation: The hidden malware. *IEEE Security Privacy*, 9(5):41–47, 2011.
40. Daniel Park, Haidar Khan, and Bülent Yener. Generation & evaluation of adversarial examples for malware obfuscation. In *2019 18th IEEE International Conference on Machine Learning and Applications (ICMLA)*, pp. 1283–1290, Boca Raton, FL, 2019.
41. PyTorch. Company Website: https://pytorch.org/tutorials/beginner/dcgan˙faces˙tutorial.html.
42. Alec Radford, Luke Metz, and Soumith Chintala. Unsupervised representation learning with deep convolutional generative adversarial networks. *CoRR*, abs/1511.06434, 2016.
43. Alec Radford, Luke Metz, and Soumith Chintala. Unsupervised representation learning with deep convolutional generative adversarial networks, 2016.
44. Waseem Rawat and Zenghui Wang. Deep convolutional neural networks for image classification: A comprehensive review. *Neural Computation*, 29(9):2352–2449, 2017.
45. Zhongru Ren, Haomin Wu, Qian Ning, Iftikhar Hussain, and Bingcai Chen. End-to-end malware detection for android IoT devices using deep learning. *Ad Hoc Networks*, 101:102098, 2020.
46. Satheesh Kumar Sasidharan and Ciza Thomas. A survey on metamorphic malware detection based on hidden markov model. In *2018 International Conference on Advances in Computing, Communications and Informatics (ICACCI)*, pp. 357–362, Bangalore, 2018.
47. Mohit Sewak, Sanjay K. Sahay, and Hemant Rathore. An investigation of a deep learning based malware detection system. *CoRR*, abs/1809.05888, 2018.
48. Mohit Sewak, Sanjay K. Sahay, and Hemant Rathore. Doom. *Adjunct Proceedings of the 2020 ACM International Joint Conference on Pervasive and Ubiquitous Computing and Proceedings of the 2020 ACM International Symposium on Wearable Computers*, September 2020.

49. Vishal Shah and Neha Sajnani. Multi-class image classification using cnn and tflite. *International Journal of Research in Engineering, Science and Management*, 3(11):65–68, 2020.
50. Neha Sharma, Vibhor Jain, and Anju Mishra. An analysis of convolutional neural networks for image classification. *Procedia Computer Science*, 132:377–384, 2018. *International Conference on Computational Intelligence and Data Science.*
51. Sanjay Sharma, Challa Rama Krishna, and Sanjay K. Sahay. Detection of advanced malware by machine learning techniques. *CoRR*, abs/1903.02966, 2019.
52. Abhishek Singh, Debojyoti Dutta, and Amit Saha. Migan: Malware image synthesis using gans. *Proceedings of the AAAI Conference on Artificial Intelligence*, 33(01):10033–10034, 2019.
53. Michael Spreitzenbarth, Felix Freiling, Florian Echtler, Thomas Schreck, and Johannes Hoffmann. Mobile-sandbox: Having a deeper look into android applications. In *Proceedings of the 28th Annual ACM Symposium on Applied Computing, SAC '13*, pp. 1808–1815, New York, NY, 2013. Association for Computing Machinery.
54. Basant Subba and Prakriti Gupta. A tfidfvectorizer and singular value decomposition based host intrusion detection system framework for detecting anomalous system processes. *Journal of Computer Security*, 100:102084, 2021.
55. Mingdong Tang and Quan Qian. Dynamic API call sequence visualisation for malware classification. *IET Information Security*, 13(4):367–377, 2019.
56. Alaa Tharwat. Classification assessment methods. *Applied Computing and Informatics*, 17(1):168–192, 2021.
57. C. George Thomas. *Research Methodology and Scientific Writing 2nd Edition*. Springer Nature, London, 2020.
58. P. Vinod, Akka Zemmari, and Mauro Conti. A machine learning based approach to detect malicious android apps using discriminant system calls. *Future Generation Computer Systems*, 94:333–350, 2018.
59. Corrado Aaron Visaggio. Static analysis for the detection of metamorphic computer viruses using repeated-instructions counting heuristics. *Journal of Computer Virology and Hacking Techniques*, 10(1):11–27, 2013.
60. Fei Xiao, Zhaowen Lin, Yi Sun, and Yan Ma. Malware detection based on deep learning of behavior graphs. *Mathematical Problems in Engineering*, 2019:8195395, 2019.
61. Yang Xin, Lingshuang Kong, Zhi Liu, Yuling Chen, Yanmiao Li, Hongliang Zhu, Mingcheng Gao, Haixia Hou, and Chunhua Wang. Machine learning and deep learning methods for cybersecurity. *IEEE Access*, 6:35365–35381, 2018.
62. Ilsun You and Kangbin Yim. Malware obfuscation techniques: A brief survey. In *2010 International Conference on Broadband, Wireless Computing, Communication and Applications*, pp. 297–300, Fukuoka, 2010.
63. Bo Yu, Ying Fang, Qiang Yang, Yong Tang, and Liu Liu. A survey of malware behavior description and analysis. *Frontiers of Information Technology & Electronic Engineering*, 19:583–603, 2018.
64. Lan Zhang, Peng Liu, and Yoon-Ho Choi. Semantic-preserving reinforcement learning attack against graph neural networks for malware detection. *CoRR*, abs/2009.05602, 2020.
65. Xuetao Zhang, Jinshuang Wang, Meng Sun, and Yao Feng. *AndrOpGAN: An Opcode GAN for Android Malware Obfuscations*, pp. 12–25, 2020.

66. Shuang Zhao, Xiaobo Ma, Wei Zou, and Bo Bai. DeepCG: Classifying metamorphic malware through deep learning of call graphs, pp. 171–190, 2019.
67. Fangtian Zhong, Xiuzhen Cheng, Dongxiao Yu, Bei Gong, Shuaiwen Song, and Jiguo Yu. MalFox: Camouflaged adversarial malware example generation based on C-GANs against black-box detectors. *CoRR*, abs/2011.01509, 2020.
68. Fangtian Zhong, Xiuzhen Cheng, Dongxiao Yu, Bei Gong, Shuaiwen Song, and Jiguo Yu. MalFox: Camouflaged adversarial malware example generation based on conv-gans against black-box detectors, 2021.

3 A Physical-Layer Approach for IoT Information Security During Interference Attacks

Abdallah Farraj
Texas A&M University - Texarkana

Eman Hammad
Texas A&M University

3.1 INTRODUCTION

The Internet of Things (IoT) is defined as the network of physical devices or "things" that are embedded with electronics, software, different kinds of sensors, and actuators and are connected to the internet via heterogeneous access networks to enable "things" to exchange data with the manufacturer, operator, and/or other connected devices [22,27]. Industrial IoT (IIoT) is a specialized IoT device that is designed as part of industrial processes or products. Considering communication requirements, IIoT can be classified into three categories: sensors that mainly transmit measurements, actuators that mainly receive control commands, and sensors/actuators that combine the capabilities to transmit and receive. IIoT industrial use cases are vast, where they can perform sensing and actuation tasks with minimal human intervention [16,22].

Enabled by innovative technologies such as 5G/6G wireless connectivity, artificial intelligence, and machine learning, IoT will continue to find enormous opportunities in applications across a wide range of industry verticals. IoT is being widely deployed in industries such as healthcare, energy, transportation, and manufacturing, to name a few [15,23,24,29]. These innovations are motivating a massive IoT adoption trend that predicts the connectivity of 75.44 billion devices by 2025 [29]. The increased utilization of IoT in critical and sensitive processes underscores the need to establish strong controls to ensure trusted and reliable operation. From a communication perspective, large-scale deployments of IoT can be supported by massive machine type communication (MTC) and machine-to-machine links; however, security might not be trivial at such scales.

DOI: 10.1201/9781003199410-3

Information security describes the technologies and practices used to support the availability, integrity, and confidentiality of information where expected. Security is typically accomplished through measures including prevention/protection, detection, and recovery. Prevention focuses on protecting information assets and services against unauthorized access, disruption, and modification. Information confidentiality maps to information not being disclosed to unauthorized parties, and integrity emphasizes maintaining accurate and complete data where unauthorized parties cannot edit the data. Availability indicates that the information is available when needed by the legitimate user or service.

Security is a critical element in the design and use of wireless networks due to the easy access of the shared radio communication channel. This is further complicated when considering IoT ecosystems, because of emerging complex cyber threats and because of the limited resources on the IoT device level. The limited resources often cause the implementation of traditional security controls, such as encryption and key management, to be impractical [25,32]. This is particularly true for most IoT devices that have low storage and processing powers with limited transmission capabilities [17].

The same characteristics that make IoT appealing for a wide range of applications present a challenge for vendors and operators to secure IoT. An IoT ecosystem can be exploited by an adversary through the IoT device hardware, communication channels & protocols, applications, and software. Hence, traditional security approaches on the application and network layers such as encryption and key management schemes can be impractical considering the IoT's light computational capabilities [32]. Without proper security controls, the benefits of IoT ecosystems cannot be realized for operational excellence, and significant damage could be inflected where IoT devices are part of closed-loop critical industrial and operational processes [17]. In these environments, the security requirements of information availability and integrity are of utmost importance. Availability ensures that controllers have timely and reliable access to IoT data when needed. However, information availability can be negatively impacted by adversarial activities or operational issues including cyberattacks, failures, and human errors.

Most IoT devices are manufactured with a type of wireless connectivity (cellular, WiFi, LPWA, etc.). Interestingly, in a wireless communication environment, adversarial actions that could degrade the communication channel are of serious concern, specifically in critical industrial settings. For example, cyberattacks that target availability include denial of service (DoS) attacks and jamming attacks. Those attacks prevent the timely exchange of legitimate IoT data with intended destinations and negatively impact the operations of the controlled system. Recent attacks, such as the Mirai attack, emphasize that resource-constrained IoT systems employing devices with limited computational and storage capabilities are vulnerable to cyber threats [2,5,17].

3.2 CHAPTER CONTRIBUTIONS

This chapter acknowledges the IoT security challenges stemming from the resource and capability limitations and adopts an alternative promising "physical-layer" security approach that takes into consideration such limited resources. We consider a legitimate IoT device communicating with a receiver unit over a wireless channel. The communication channel is modeled to be under an attack by an *Adversary* by means of an intentional jamming interference, which affects the quality of the received signal at the receiver unit, as is illustrated in Figure 3.1.

We employ a cross-layer physical-security approach by focusing on the quality of service (QoS) of the received signal while meeting the IoT power constraints. Specifically, in the proposed system setup, the IoT device would want to remain within a "specific" acceptable outage limit while constraining the number of retransmissions. Packet retransmissions might be needed in response to channel conditions and the adversarial jamming interference.

In this setup, the IoT device cannot control the transmission strategy of the adversary, but it can control its own strategy in order to conserve limited resources. Hence, the IoT device evolves its own strategy without adding any additional coordination overhead, which is particularly important in large-scale IoT deployments.

To the best of our knowledge, this approach to managing jamming interference while meeting target information availability QoS has not been addressed before. The main contributions of this chapter can be summarized as follows:

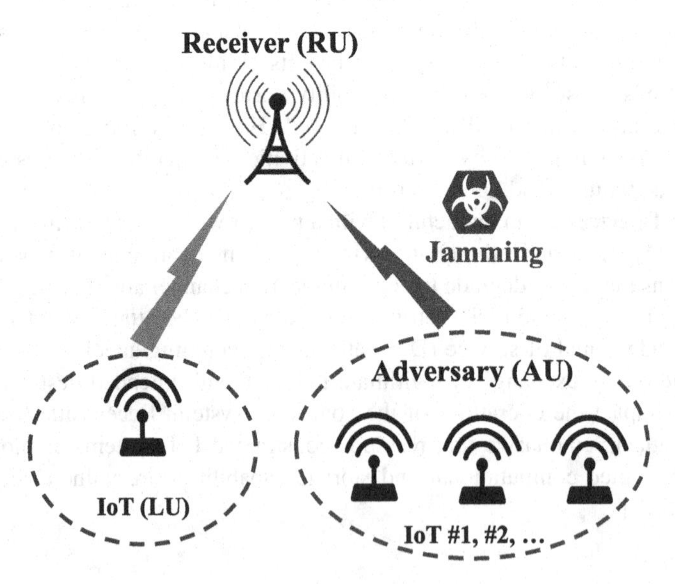

Figure 3.1 Problem setup showing an adversary jamming the communication of the legitimate IoT device.

- Develop a model for wireless heterogeneous environment with a setup of IoT devices (wireless connectivity) competing for channel access in the presence of adversarial IoT devices inducing jamming attacks.
- Develop channel-access strategies that meet QoS metrics and security objective (specifically, information availability) while considering the limited resources of the IoT device.
- Demonstrate the benefits of the proposed approach using analytical and numerical simulation results.

3.3 RELATED WORK

In this work, we combine several approaches such as physical-layer security, cognitive spectrum sharing, and scheduling to establish practical and scalable security strategies for IoT. Existing literature started by extending security approaches from sensor networks to IoT for obvious similarities [4,14,15,18,19,25,30,31,33]. The work of Ref. [25] provides a review of physical-layer security approaches to achieve confidentiality over wireless channels.*Physical-layer security* utilizes the physics of wave propagation and transceiver design to enable some aspects of secure communications over the wireless channel [4,22,25]. Hence, physical-layer security has a promising potential to overcome the limitations of traditional security measures for IoT applications.

Spectrum-sharing cognitive technologies are evolving as an enabler of IoT connectivity in newer generations of wireless communication systems [6,7]. Device scheduling has been typically addressed through centralized user scheduling schemes [8,13]. However, the projected massive scale of IoT deployments and the emerging different characteristics (e.g., energy efficiency requirements) are challenging to existing security approaches of sensor networks. For example, probabilistic ciphering, compressive sensing, and approaches dependent on channel state information (CSI) have been shown to not scale well, while some (e.g., compressive sensing) have impractical high-computational complexity [22,25].

Recent works that investigate the security of IoT systems include [3,20,21,28]. The work in Ref. [3] investigates the architecture and techniques for IoT security and privacy. Further, physical-layer security approaches to IoT information availability and confidentiality are considered in Refs. [9,10]; these works utilized employing other IoT devices in the system to safeguard against interference attacks or eavesdroppers that target a specific IoT device.

3.4 IoT INFORMATION SECURITY

We adopt zero-determinant games to devise an uncoordinated transmission strategy for an IoT device experiencing jamming interference. The strategy's objective is to achieve a target QoS metric (specifically, outage probability) given the limited resources of the IoT device (manifested in reducing the average number of channel transmissions).

3.4.1 BACKGROUND

In this chapter, we focus on one legitimate user and one adversary user of the channel to enable tractable treatment for the reader. However, we emphasis that the zero-determinant games framework adopted here is easily extendable to multiple users, which makes it appropriate to model large-scale IoT deployments.

A dynamical model is developed to capture the impact of uncoordinated transmissions over the wireless channel, and then the impact of the IoT device's transmission strategy on signal quality is investigated. The dynamic interactions between the IoT device and the adversary are modeled as a 2×2 iterated game; in each round of the game, the IoT device reacts to the actions of the adversary in the previous round. In the presence of an ongoing jamming interference created by the adversary, the proposed game-theoretic transmission strategy allows the IoT device to achieve a target outage probability as the information availability QoS metric without additional coordination overhead.

An important feature of the proposed game-theoretic transmission strategy is that players with longer memories of the game history have no advantage, in the long term, over those with shorter memories. These iterated games are referred to as *zero-determinant* strategies [11,12,26], and players can control their own long-term payoff or that of their opponent through the structure of the game payoff matrix.

Thus, to be able to use this strategy, the IoT device does not need to know the history of the adversary's transmission; also, the IoT device can achieve, on the long run, a target performance given the structure of the payoff matrix. Consequently, the IoT device will be able to achieve the QoS target while transmitting over the wireless channel without the overhead of coordinating with a scheduling authority or other users of the channel.

3.4.2 SYSTEM MODEL

Consider a generic wireless communications setup as illustrated in Figure 3.2. The system includes a legitimate user of the system (termed as LU) that uses the wireless channel to transmit its data a receiver unit (denoted RU). The system includes an adversary user (termed AU) that causes jamming interference on the LU's signal. Both LU and AU devices can concurrently transmit over the wireless channel.

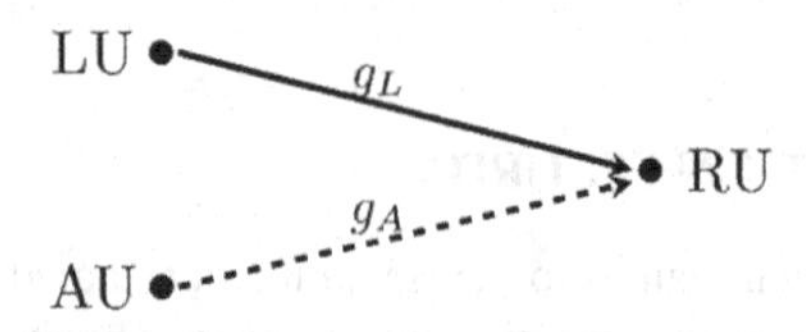

Figure 3.2 System model displaying the channel power gain between the legitimate user (AU) and the receiver unit (RU) and the channel power gain between the adversary user (AU) and RU.

Further, LU is assumed to have a transmission rate of R_L and a transmission power of P_L that is constant over the time period of interest. Similarly, AU has a constant transmission power of P_A and a transmission rate of R_A. The noise at the receiver unit is assumed to be an additive white Gaussian noise with zero mean and variance of σ^2.

The channels between the two devices (LU, AU) and the receiver unit (RU) are assumed to be independent and identically distributed (i.i.d.) block-fading channels with Rayleigh distribution. The channel power gain between LU and RU is denoted as g_L, and the channel power gain between AU and RU is termed as g_A. Let γ_L denotes the signal-to-interference plus noise ratio (SINR) of LU's signal at RU, and let $\mathbb{P}\{\cdot\}$ be the probability operator. Thus, the outage probability of LU's transmission is $\mathbb{P}\{\log_2(1+\gamma_L) \leq R_L\}$.

Consider the case when there is no interference over the wireless channel (i.e., $P_A = 0$). The signal-to-noise ratio (SNR) of LU's signal at RU is $\gamma_L = \frac{P_L g_l}{\sigma^2}$. The outage probability of LU (denoted ζ_0 in this case) is calculated using $\zeta_0 = \mathbb{P}\{\log_2(1+\frac{P_L g_l}{\sigma^2}) \leq R_A\}$. Thus,

$$\zeta_0 = 1 - \exp(\frac{-\sigma^2}{P_L}(2^{R_L}-1)) \tag{3.1}$$

for Rayleigh channel. Next, consider LU's outage probability under the interference caused by AU. In this case,

$$\gamma_L = \frac{P_L g_L}{P_A g_A + \sigma^2}. \tag{3.2}$$

The probability density function (PDF) of γ_L can be found as:

$$f_{\gamma_L}(x) = \frac{P_L}{P_A}\exp(-\frac{\sigma^2}{P_L}x)\frac{1}{(x+\frac{P_L}{P_A})^2}\left(\frac{\sigma^2}{P_L}(x+\frac{P_L}{P_A})+1\right), x \geq 0. \tag{3.3}$$

The outage probability of LU during interference is calculated from $\zeta_L = \mathbb{P}\{\gamma_L \leq 2^{R_L}-1\}$ as:

$$\begin{aligned}\zeta_L &= 1 - \frac{1-\zeta_0}{1+\frac{P_A}{P_L}(2^{R_L}-1)} \\ &= 1 - \frac{\exp(\frac{-\sigma^2}{P_L}(2^{R_L}-1))}{1+\frac{P_A}{P_L}(2^{R_L}-1)}.\end{aligned} \tag{3.4}$$

3.5 ZERO-DETERMINANT STRATEGIES

Consider a 2×2 iterated game with the one stage game of Table 3.1. The game has two players: User 1 (row player) and User 2 (column player). At each round of the game, a player chooses from two actions $\{1,2\}$. Let n_1 and n_2 denote the actions of User 1 and User 2, respectively. A value of $n_1 = 1$ or $n_1 = 2$ refers to an active or

Table 3.1
Generic Payoff Matrix

User 1 \ User 2	$n_2 = 1$ (active)	$n_2 = 2$ (idle)
$n_1 = 1$ (active)	$X_{1,1}$	$X_{1,2}$
$n_1 = 2$ (idle)	$X_{2,1}$	$X_{2,2}$

idle User 1 in a given round of the game, respectively, and a similar interpretation is used for $n_2 = 1$ (active User 2) and $n_2 = 2$ (idle User 2). The value $X_{j,k}$, where $j,k \in \{1,2\}$, denotes the game payoff if User 1 chooses $n_1 = j$ and User 2 chooses $n_2 = k$ during the current play interval.

It is shown in Ref. [26] that in iterated games, where the same actions and the same payoff matrices are repeated, for any strategy of the player with the longer memory, the player with the shorter memory can achieve the same long-term outcome if the opponent has played a shorter memory strategy. Thus, any history outside what is shared between the two players can be disregarded. Consequently, the game can be modeled as a Markov chain taken here to be a single memory step process.

In this regard, let $\boldsymbol{n}(t) = (n_1, n_2)$ denote the state of the game at round t and $\boldsymbol{S} = \{(1,1),(1,2),(2,1),(2,2)\}$ be the state space of the game. Also let $\boldsymbol{k} = (k_1, k_2)$, then

$$p_1^{\boldsymbol{k}} = \mathbb{P}(n_1(t+1) = 1 \mid \boldsymbol{n}(t) = \boldsymbol{k}), \forall \boldsymbol{k} \in \boldsymbol{S} \tag{3.5}$$

represents the probability that User 1 takes action 1 ($n_1 = 1$) in round $t+1$ if in the previous round User 1 took action k_1 and User 2 took action k_2. In a similar manner, the probability that User 2 takes action 1 ($n_2 = 1$) in round $t+1$ is represented as:

$$p_2^{\boldsymbol{k}} = \mathbb{P}(n_2(t+1) = 1 \mid \boldsymbol{n}(t) = \boldsymbol{k}), \forall \boldsymbol{k} \in \boldsymbol{S}. \tag{3.6}$$

The Markov chain has a unique stationary distribution $\pi^T = [\pi_{1,1}, \pi_{1,2}, \pi_{2,1}, \pi_{2,2}]$, where $\pi_{j,k}$, $\forall j,k \in \{1,2\}$, refers to the stationary probability that User 1 takes action j and User 2 takes action k. The average long-term outcome of the game of the row player, denoted as u_1, is given by Ref. [1] as:

$$u_1 = \pi^T \hat{\boldsymbol{X}} \tag{3.7}$$

where $\hat{\boldsymbol{X}} = [X_{1,1}, X_{1,2}, X_{2,1}, X_{2,2}]^T$. It is shown in Ref. [26] that if the $p_1^{\boldsymbol{k}}$'s are chosen such that

$$a_1 \hat{\boldsymbol{X}} + b_1 = \left[-1 + p_1^{1,1}, -1 + p_1^{1,2}, p_1^{2,1}, p_1^{2,2}\right]^T \tag{3.8}$$

where a_1 and b_1 are arbitrary nonzero real numbers, then the row player can fix the value of u_1 regardless of the actions of the column player (User 2) if and only if the minimum value of one row in the payoff matrix of Table 3.1 exceeds the maximum value of the other row. In such case, the row player (i.e., User 1) can fix the long-term average payoff u_1 to any value in the range between the minimum and the maximum values representing an advantage to User 1 over the other user [1].

To achieve a specific long-term average payoff u_1, User 1 has to take an action in the current interval according to the following likelihoods [1]:

$$\begin{aligned} p_1^{1,1} &= 1 + \left(1 - \frac{X_{1,1}}{u_1}\right) b_1 \\ p_1^{1,2} &= 1 + \left(1 - \frac{X_{1,2}}{u_1}\right) b_1 \\ p_1^{2,1} &= \left(1 - \frac{X_{2,1}}{u_1}\right) b_1 \\ p_1^{2,2} &= \left(1 - \frac{X_{2,2}}{u_1}\right) b_1 . \end{aligned} \tag{3.9}$$

The range of valid values of b_1 can be found as Ref. [1]:

$$0 < b_1 \leq \min\left(\frac{-1}{1 - \frac{X_{1,\max}}{u_1}}, \frac{1}{1 - \frac{X_{2,\min}}{u_1}}\right). \tag{3.10}$$

3.6 GAME-THEORETIC TRANSMISSION STRATEGY

The objective of the legitimate IoT device LU is to transmit its own data over the shared wireless channel to the intended receiver unit RU while subject to the jamming interference created by the adversary user AU. The information availability is one objective the IoT device has to achieve, which can be quantified by using the outage probability of LU's signals as the QoS metric of interest. Because of the relationship between outage probability and the corresponding SINR value, γ_L will be used as the payoff of a transmission period.

Let $\boldsymbol{X} = \left[X_{j,k}\right]$ denotes the payoff matrix of the legitimate IoT device LU during transmission period ΔT. Then, the values of $\boldsymbol{X}$ are shown in Table 3.2, where the term *Active* indicates the device is transmitting over the wireless channel during ΔT, and *Idle* means no transmission by the device.

3.6.1 TRANSMISSION PROBABILITY

Let User 1 in the zero-determinant game framework represent LU since it is the IoT device of interest; also, let User 2 represent the adversary (AU). LU's goal is to trans-

Table 3.2
Payoff Matrix of LU

LU \ AU	$n_2 = 1$ (active)	$n_2 = 2$ (idle)
$n_1 = 1$ (active)	$\frac{P_L g_L}{P_A g_A + \sigma^2}$	$\frac{P_L g_L}{\sigma^2}$
$n_1 = 2$ (idle)	0	0

mit its data over the channel to RU while meeting an outage probability requirement (as an information availability measure).

Because the outage probability of LU is $\mathbb{P}\{\log_2(1+\gamma_L) \leq R_L\}$, we use the value of the SINR (γ_L) to represent the *payoff* at the end of the time interval ΔT for LU. Thus, as shown in Table 3.2, the payoff matrix for LU is:

$$\boldsymbol{X} = \begin{bmatrix} \frac{P_L g_L}{P_A g_A + \sigma^2} & \frac{P_L g_L}{\sigma^2} \\ 0 & 0 \end{bmatrix}. \tag{3.11}$$

Let $X_{j,\max}$ and $X_{j,\min}$ denote the maximum and minimum values of row j in the payoff matrix of LU, respectively. Such values are found for each row as:

$$\begin{aligned} X_{1,\min} &= X_{1,1} = \frac{P_L g_L}{P_A g_A + \sigma^2} \\ X_{1,\max} &= X_{1,2} = \frac{P_L g_L}{\sigma^2} \\ X_{2,\min} &= X_{2,1} = 0 \\ X_{2,\max} &= X_{2,2} = 0. \end{aligned} \tag{3.12}$$

It is observed that $X_{2,\max} < X_{1,\min}$. Thus, the average long-term payoff attained by LU (termed, u_L), using the zero-determinant strategy, lies in $[X_{2,2}, X_{1,1}]$, or

$$u_L \in \left[0, \frac{P_L g_L}{P_A g_A + \sigma^2}\right]. \tag{3.13}$$

Let the specific value of the average long-term payoff (in other words, SINR) attained by LU be parameterized as:

$$u_L = \alpha_L \frac{P_L g_L}{P_A g_A + \sigma^2} \tag{3.14}$$

where $0 < \alpha_L \leq 1$ is called the *proactiveness factor* of LU. A high value of α_L moves u_L closer to $X_{1,1}$, indicating that LU is more aggressive in utilizing the wireless channel to achieve higher SINR values and thus lower outage probability. The case of $\alpha_L = 0$ indicates the lack of transmission of LU, and it is not of practical interest in this chapter.

Further, following the development in Ref. [1], the range of valid values of b_1 in Eqs. (3.8) and (3.9) is found as:

$$0 < b_1 \leq \min\left(\frac{u_L}{X_{1,\max} - u_L}, \frac{u_L}{u_L - X_{2,\min}}\right). \tag{3.15}$$

This leads to $0 < b_1 \leq \min\left(\frac{u_L}{X_{1,\max} - u_L}, 1\right)$. Also, define:

$$b_{1,\max} = \begin{cases} 1 & \alpha_L \geq \frac{P_A g_A}{2\sigma^2} + \frac{1}{2} \\ \frac{\alpha_L \sigma^2}{P_A g_A + (1-\alpha_L)\sigma^2} & \alpha_L < \frac{P_A g_A}{2\sigma^2} + \frac{1}{2} \end{cases}. \tag{3.16}$$

Then, the specific value of b_1 can be expressed as:

$$b_1 = \beta_L b_{1,\max} \tag{3.17}$$

where $0 < \beta_L \leq 1$ is called the *reactiveness factor* of LU. A high value of β_L means that LU is more probable to take an *Active* action if it was idle in the previous ΔT interval, and it is less probable to transmit in the current interval if it did in the previous one.

Let the status of the IoT device (LU) and the adversary (AU) in the previous time interval ΔT be j and k, respectively, where $j,k \in \{1,2\}$, and let k be known to LU. Then, $p_1^{j,k}$ denotes the probability that LU is active in the current time interval given its knowledge that $n_1 = j$ and $n_2 = k$ in the previous transmission interval.

Using the results of Eqs. (3.9) and (3.14), the probabilities of LU transmitting over the wireless channel are:

$$\begin{aligned} p_1^{1,1} &= 1 + \tfrac{u_L - X_{1,1}}{u_L} b_1 \\ &= 1 - \tfrac{1-\alpha_L}{\alpha_L} b_1 \\ p_1^{1,2} &= 1 + \tfrac{u_L - X_{1,2}}{u_L} b_1 \\ &= p_1^{1,1} - \tfrac{P_A g_A}{\alpha_L \sigma^2} b_1 \\ p_1^{2,1} &= \tfrac{u_L - X_{2,1}}{u_L} b_1 \\ &= b_1 \\ p_1^{2,2} &= \tfrac{u_L - X_{2,2}}{u_L} b_1 \\ &= b_1 \,. \end{aligned} \tag{3.18}$$

Consider the case of $\alpha_L = 1$ and a jamming interference higher than the noise power (i.e., $P_A g_A > \sigma^2$), then Eq. (3.18) reduces into:

$$\begin{aligned} p_1^{1,1} &= 1 \\ p_1^{1,2} &= 1 - \beta_L \\ p_1^{2,1} &= p_1^{2,2} = \tfrac{\sigma^2}{P_A g_A} \beta_L < \beta_L \,. \end{aligned} \tag{3.19}$$

Similarly, when the jamming interference power is lower than that of the noise, Eq. (3.18) is simplified to:

$$\begin{aligned} p_1^{1,1} &= 1 \\ p_1^{1,2} &= 1 - \tfrac{P_A g_A}{\sigma^2} \beta_L > 1 - \beta_L \\ p_1^{2,1} &= p_1^{2,2} = \beta_L \,. \end{aligned} \tag{3.20}$$

3.6.2 TRANSMISSION STRATEGY

Let $\boldsymbol{M}$ denote the state transition matrix of the Markov chain. Thus,

$$\boldsymbol{M} = \begin{bmatrix} p_1^{1,1} p_2^{1,1} & p_1^{1,1}(1-p_2^{1,1}) & (1-p_1^{1,1})p_2^{1,1} & (1-p_1^{1,1})(1-p_2^{1,1}) \\ p_1^{1,2} p_2^{2,1} & p_1^{1,2}(1-p_2^{2,1}) & (1-p_1^{1,2})p_2^{2,1} & (1-p_1^{1,2})(1-p_2^{2,1}) \\ p_1^{2,1} p_2^{1,2} & p_1^{2,1}(1-p_2^{1,2}) & (1-p_1^{2,1})p_2^{1,2} & (1-p_1^{2,1})(1-p_2^{1,2}) \\ p_1^{2,2} p_2^{2,2} & p_1^{2,2}(1-p_2^{2,2}) & (1-p_1^{2,2})p_2^{2,2} & (1-p_1^{2,2})(1-p_2^{2,2}) \end{bmatrix} \tag{3.21}$$

where the p_1's are defined in Eq. (3.18) and the p_2's represent the adversary's transmission strategy (which is not necessarily known to the IoT device). Following Eq. (3.7), $\pi_{j,k}, \forall j,k \in \{1,2\}$, denotes the stationary probability that LU takes action j and AU takes action k. The stationary distribution π is calculated as Refs. [1,26]

$$\pi^T = [\pi_{1,1}, \pi_{1,2}, \pi_{2,1}, \pi_{2,2}] = \pi^T \boldsymbol{M}. \tag{3.22}$$

Thus, π can be obtained using a normalized version of the left eigenvector of $\boldsymbol{M}$ with a corresponding eigenvalue of 1.

The stationary transmission probability of LU is $\pi_{1,1} + \pi_{1,2}$, and the mean long-term transmission power of LU is then calculated using:

$$\overline{P_L} = (\pi_{1,1} + \pi_{1,2}) P_L. \tag{3.23}$$

Since LU is resource constrained, let $\overline{P_L} \leq P_{1,\max}$. Further, the mean long-term bandwidth-normalized channel capacity achieved by LU ($\overline{C_L}$) is found as:

$$\overline{C_L} = \frac{\pi_{1,1} \log_2 (1 + \frac{P_L g_L}{P_A g_A + \sigma^2}) + \pi_{1,2} \log_2 (1 + \frac{P_L g_L}{\sigma^2})}{\pi_{1,1} + \pi_{1,2}}. \tag{3.24}$$

Similarly, the average long-term outage probability of LU that results from applying the game-theoretic transmission strategy is found as:

$$\overline{\zeta_L} = \frac{\pi_{1,1} \zeta_L + \pi_{1,2} \zeta_0}{\pi_{1,1} + \pi_{1,2}} \tag{3.25}$$

where ζ_0 and ζ_L are defined in Eqs. (3.1) and (3.4), respectively.

The IoT device's strategy is to achieve one objective of information security through guaranteeing a level of information availability; this goal is accomplished by limiting the outage probability to some threshold. Through the use of a specific value of α_L, LU can, on the long term, achieve a target SINR value, which is translated into a target outage probability. Further, the IoT device conserves its transmission power as it does not have to transmit all the time in order to achieve this QoS measure by following the results in Eq. (3.18).

Since the IoT device's channel access depends on the adversary strategy, the IoT device can utilize the value of β_L from Eq. (3.17) to select the transmission probabilities and eventually the stationary *Active* probability of Eq. (3.22). As a summary, the IoT device:

- selects the value of α_L to guarantee long-term performance (u_L and $\overline{\zeta_L}$),
- selects the value of β_L to meet resource-usage constraint ($\overline{P_L} \leq P_{1,\max}$).

Algorithm 1 is a depiction of the zero-determinant strategy that LU will follow to transmit over the wireless channel while maintaining a long-term average payoff (u_L) that meets its QoS requirement regardless of the interference activity caused by AU.

Algorithm 1 Zero-Determinant Transmission Strategy

Determine: $\overline{\zeta_L}$ from Eq. (3.25) to meet QoS requirement.
Collect: $P_L, R_L, g_L, \sigma^2, P_A, g_A$.
Calculate: X from Table 3.2.
Calculate: u_L from Eq. (3.14).
Determine: α_L and β_L.
Calculate: p_1 from Eq. (3.18) for *Active, Idle* probabilities.
Initialize: $j = 2$, to denote idle status of LU in the previous ΔT.
while TRUE **do**
  Assign: $k \in \{1, 2\}$ depending on the status of AU in the previous ΔT.
  Determine: $p_1^{j,k}$.
  Generate: random number p.
  if $p_1^{j,k} \geq p$ **then**
    Access: LU transmits data with P_L over cognitive channel.
    Update: $j = 1$.
  else
    Idle: LU does not transmit.
    Update: $j = 2$.
  end if
  Find: Active/Idle status of AU in the current ΔT.
  if LU has no more data to transmit **then**
    Break Loop.
  end if
end while

3.7 EXTENSION TO MULTIPLE IoT USERS

The analysis presented for two players is next extended a repeated game with multiple players. This is necessary to examine the scalability and generality of the presented model.

3.7.1 ZERO-DETERMINANT STRATEGIES

Let $\{1, \ldots, N\}$ be the index of the game players where $N \geq 2$ is the number of players. Let $\boldsymbol{n}(t) = [n_1(t), \ldots, n_N(t)]$ denote the state of the game at round t, where $n_i(t) \in \{1, 2\}$ describes the *Active* or *Idle* binary actions $\forall t, i \in \{1, \ldots, N\}$. A multi-dimensional Markov chain can be used to describe the process $\{\boldsymbol{n}(t) : t = 0, 1, \ldots\}$, and the state transition matrix, $\boldsymbol{M}$, can be presented using a $2^N \times 2^N$ matrix.

Similar to the 2-player game, a player i in an N-player game takes a specific action in a given round with probability that depends on the state of the game in the previous round. Let $p_i^{\boldsymbol{k}}$ be the probability that player $i, \forall i = \{1, \ldots, N\}$, takes action 1 in a given round if the game was in state $\boldsymbol{k}$ in the previous round.

Further, let $\boldsymbol{X}_i^{\boldsymbol{k}}$ be the payoff of player i if the state of the game at the previous round is $\boldsymbol{k}$. Also, for $k \in \{1, 2\}$, define $\boldsymbol{X}_{i,\min}^k = \min(\boldsymbol{X}_i^{\boldsymbol{k}} : n_i = k)$ and $\boldsymbol{X}_{i,\max}^k = \max(\boldsymbol{X}_i^{\boldsymbol{k}} : n_i = k)$ as the minimum and maximum payoffs of player i when taking action k, respectively.

Regardless of the actions of the other players in the game, player i can control its long-term average payoff if $k_{i,\max}, k_{i,\min} \in \{1, 2\}$ exist such that $\boldsymbol{X}_{i,\max}^{k_{\max}} \leq \boldsymbol{X}_{i,\min}^{k_{\min}}$ [1].

If this is the case, then the long-term payoff of player i, termed u_i, can be any value in the interval $[X_{i,\max}^{k_{\max}}, X_{i,\min}^{k_{\min}}]$, and this long-term payoff can be achieved using the strategy of:

$$p_i^k = 1 + \frac{b_i}{u_i}(u_i - X_i^k) \tag{3.26}$$

as the probability of choosing action 1 when the state of the game is $\boldsymbol{k}$ where b_i depends on the value of $k_{i,\max}$ [1].

3.7.2 GENERALIZED TRANSMISSION STRATEGY

Given the N IoT in the communication system and for User 1 being the legitimate user LU, User 1 takes actions whether to transmit over the wireless channel following the zero-determinant strategy described above. In order to meet the QoS requirements, User 1 conducts a zero-determinant transmission strategy as follows:

- Calculate the $N \times N$ payoff matrix of LU
- Verify if $X_{1,\max}^{k_{\max}} \leq X_{1,\min}^{k_{\min}}$
- Define the long-term target of LU's outage probability in the range $[X_{1,\max}^{k_{\max}}, X_{1,\min}^{k_{\min}}]$ as:

$$u_L = X_{1,\max}^{k_{\max}} + \alpha_L \left(X_{1,\min}^{k_{\min}} - X_{1,\max}^{k_{\max}} \right) \tag{3.27}$$

- Select the value of b_1 that meets the constraints of $k_{1,\max}$
- For each transmission interval ΔT:
  - Determine the previous ΔT's transmission state, $\boldsymbol{k}$
  - Determine the previous ΔT's game payoff of User 1, X_1^k
  - Transmit over the wireless channel with probability

$$p_1^k = 1 + \frac{b_1}{u_L} \left(u_L - X_1^k \right). \tag{3.28}$$

As evident from the above description, the N-user case is a natural extension of the 2-player case detailed in Algorithm 1.

3.8 NUMERICAL RESULTS

In this section, we numerically illustrate the proposed transmission strategy.

3.8.1 MODEL DYNAMICS

Consider the case of $P_L = P_A = 10\sigma^2$ and $g_L = g_A = 1$. Let $\alpha_L = 1$ to signify an aggressive LU. Figure 3.3 displays the active probabilities for LU, defined in Eqs. (3.9) and (3.18) for the proposed transmission strategy.

Figure 3.3 illustrates the relationship between the transmission probabilities of LU versus the *reactiveness factor*. The numerical results of Figure 3.3 align with the

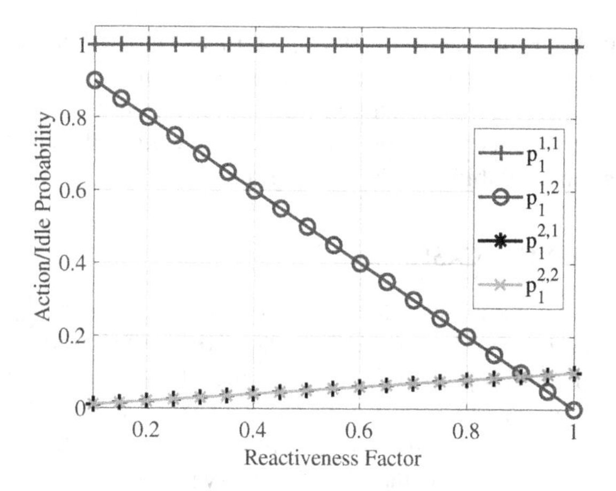

Figure 3.3 LU's *Active* probabilities versus *reactiveness factor* (β_L).

formulation deduced in Eq. (3.19), where increasing β_L leads to decreasing values of $p_1^{1,2}$, increasing those of $p_1^{2,1}, p_1^{2,2}$, and no effect on $p_1^{1,1}$.

Further, Figure 3.4 illustrates the stationary transmission probabilities of the IoT device for different values of β_L and transmission probabilities of the adversary (p_2). Equations (3.22) and (3.23) emphasize that the stationary transmission probabilities are the first two elements of the normalized eigenvector of the matrix $\boldsymbol{M}$ defined in

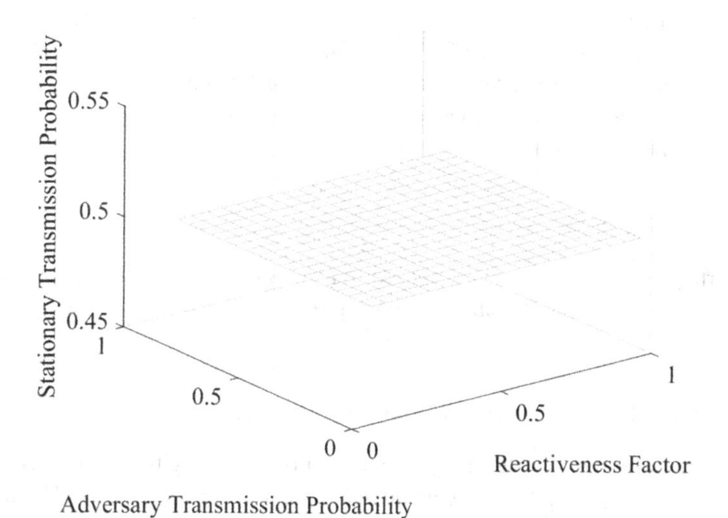

Figure 3.4 LU's stationary transmission probabilities ($\pi_{1,1} + \pi_{1,2}$) versus *reactiveness factor* (β_L) and adversary transmission probability (p_2).

Eq. (3.21). Given the values of $P_L, P_A, \sigma^2, g_L, g_A, \alpha_L$ as assigned in the previous case, $\pi_{1,1} + \pi_{1,2} = \frac{1}{2}$ as shown in this figure. This translates into a 50% chance, on the long term, of the IoT device accessing the channel and transmitting its signal; thus, the device is conserving its resources half of the time while still achieving the target level of information availability at the receiver unit.

3.8.2 SIMULATED USE CASES

Next, we simulate a communication environment with $\frac{P_L}{P_A} = 5$ dB, $\frac{P_L}{\sigma^2} = 10$ dB, and $R_L = 1$ bit/s/Hz. In this simulation, AU (i.e., the adversary) randomly transmits over the cognitive channel with probability of p_2, while LU adopts the zero-determinant transmission strategy described in Eq. (3.18) with $\alpha_L = 1, \beta_L = 0.5$. We consider four values of adversary transmission probability ($p_2 = 10\%, 40\%, 70\%, 90\%$) for Figures 3.5–3.7. We demonstrate the LU's stationary transmission probability ($\pi_{1,1} + \pi_{1,2}$), long-term average SINR (u_L), and long-term average of channel capacity ($\overline{C_L}$) in Figures 3.5–3.7, respectively.

The figures demonstrate that the stationary transmission probability increases with increasing the adversary's transmission probability (p_2); however, the IoT device does not match the increase in p_2, which provides an advantage for using the proposed zero-determinant transmission strategy.

It is also observed that the long-term average payoff (i.e., u_L) is almost the same in Figure 3.6 regardless of the adversary's strategy. This aligns with the promise of the zero-determinant transmission strategy in achieving the same payoff regardless of the actions of the other users in the system. This finding supports the IoT device's goal to limit the channel outage to a threshold and thus preserve the information availability objective. Similarly, LU achieves stable long-term outage probability and channel capacity with a wide range of aggressive interference activity by the adversary user.

Finally, the impact of number of adversary users on LU's performance metrics is shown in Figure 3.8. It is observed here that increasing the number of adversary users gracefully deteriorates the performance metrics of the legitimate user even though the channel access is uncoordinated.

3.9 DISCUSSIONS

It is important to discuss how the proposed transmission strategy fits applicable scenarios and understand the benefits and constraints of this strategy versus an active strategy.

3.9.1 ABOUT THE GAME-THEORETIC APPROACH

Recall that the average long-term payoff attained by LU using the zero-determinant strategy lies in $[X_{2,2}, X_{1,1}]$. It is to be noted that if the IoT of interest (i.e., LU) chooses to *act all the time* regardless of the actions of the adversary, the average long-term payoff will be in the range $[X_{1,1}, X_{1,2}]$. This provides LU with a better payoff than the average long-term payoff attained by using the game-theoretic approach.

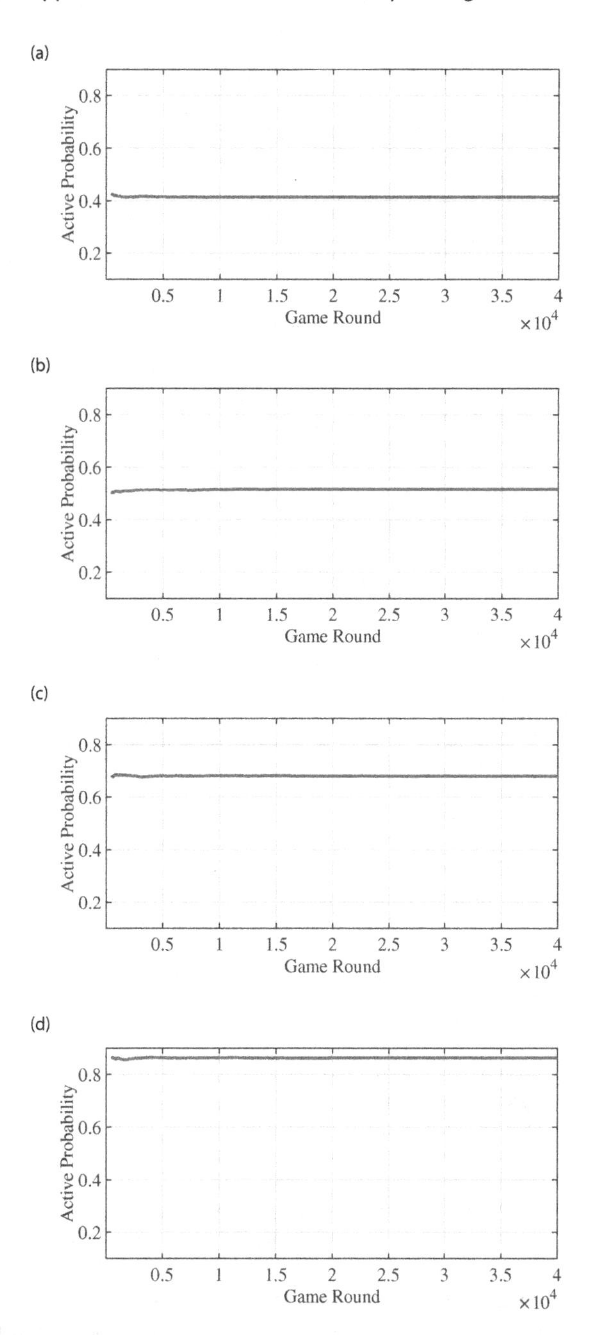

Figure 3.5 The stationary transmission probability of the legitimate user for different values of the adversary's transmission probability (p_2). (a) Stationary transmission probability when $p_2 = 10\%$. (b) Stationary transmission probability $p_2 = 40\%$. (c) Stationary transmission probability $p_2 = 70\%$. (d) Stationary transmission probability $p_2 = 90\%$.

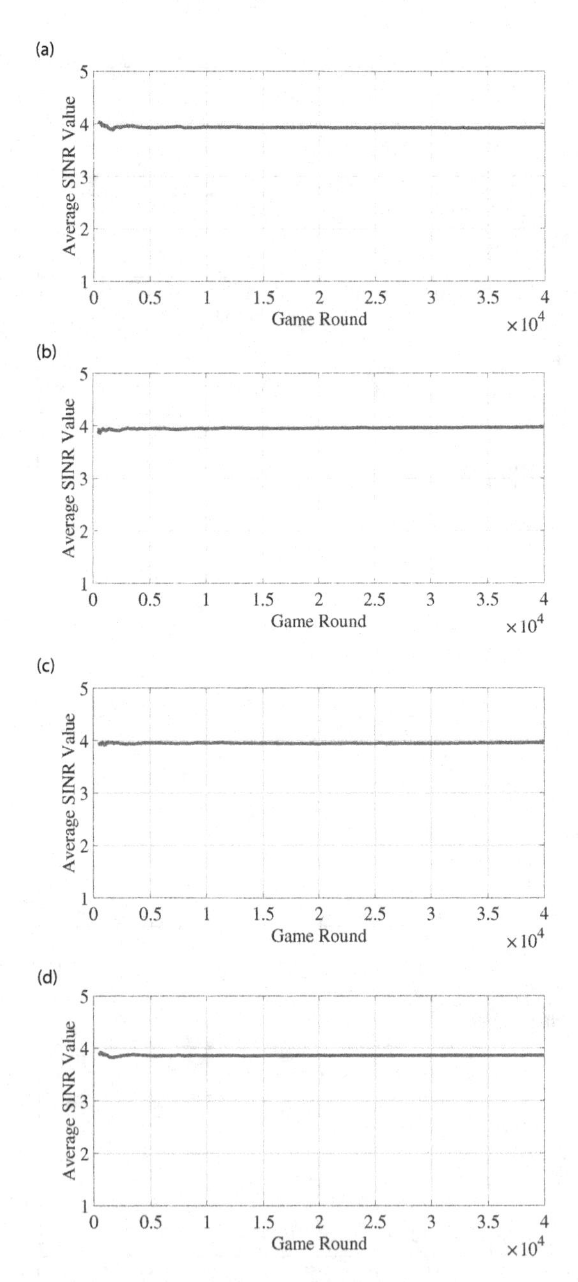

Figure 3.6 Long-term average of SNR of the legitimate user for different values of the adversary's transmission probability (p_2). (a) Long-term average of the legitimate user's SNR when $p_2 = 10\%$. (b) Long-term average of the legitimate user's SNR when $p_2 = 40\%$. (c) Long-term average of the legitimate user's SNR when $p_2 = 70\%$. (d) Long-term average of the legitimate user's SNR when $p_2 = 90\%$.

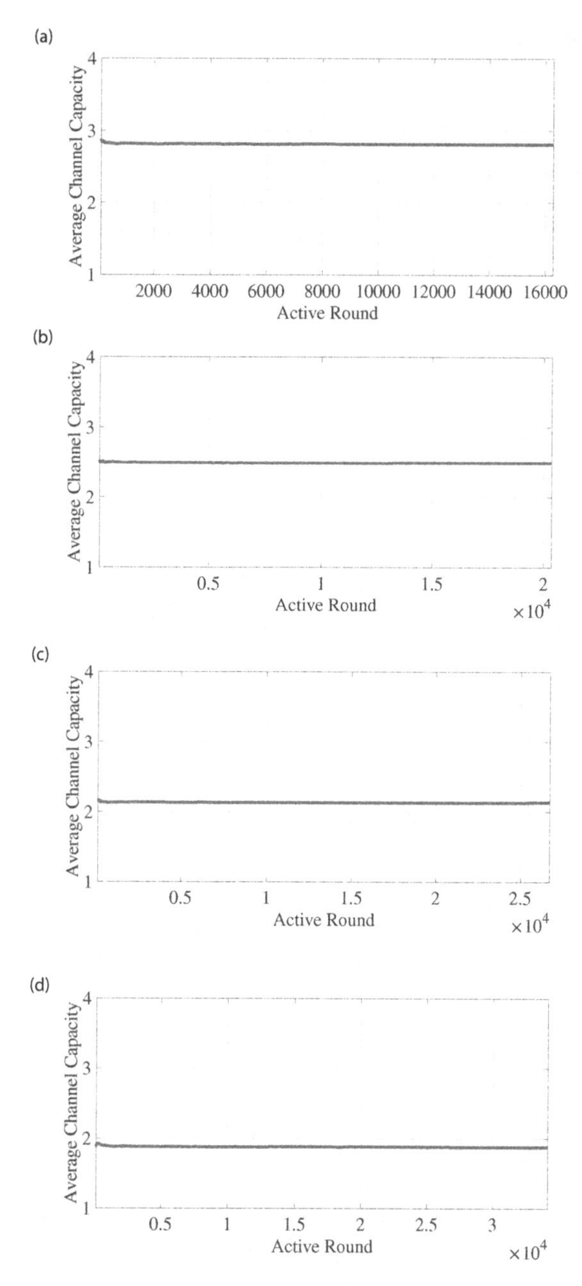

Figure 3.7 Long-term average of channel capacity of the legitimate user for different values of the adversary's transmission probability (p_2). (a) Average channel capacity of LU when $p_2 = 10\%$. (b) Average channel capacity of LU when $p_2 = 40\%$. (c) Average channel capacity of LU when $p_2 = 70\%$. (d) Average channel capacity of LU when $p_2 = 90\%$.

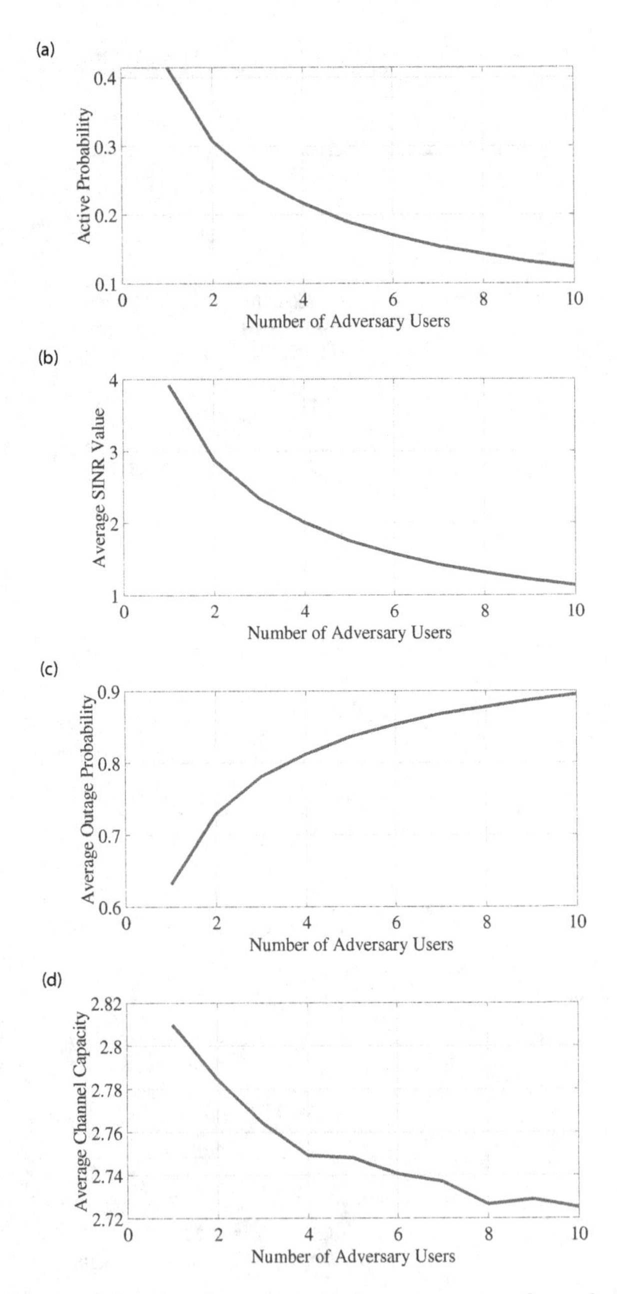

Figure 3.8 Impact of changing the number of adversary users on the performance metrics of the legitimate user of the system (LU). (a) Active probability of the legitimate user versus number of adversary users. (b) Impact of varying the number of adversary users on the average SNR value of the legitimate user. (c) Average outage probability of the legitimate user as a function of the number of adversary users. (d) Impact of number of adversary users on the average channel capacity of the legitimate user.

However, the game-theoretic control approach gives the LU a guarantee of meeting its QoS requirement while meeting constraints on the data availability or the cost of using the cognitive channel. This is illustrated by how the proposed model incorporates the limited sampling rate of IoT sensor readings as leveraged by LU. The analysis does not assume the availability of a continuous stream of data, and the bound on available cognitive resources or cost is articulated as the process of selective cognitive channel transmission.

An advantage of the proposed game-theoretic treatment is that the IoT device of interest can choose to be an active user of the wireless channel or idle to conserve resources. The legitimate user of the channel is not required to know the entire history of the adversary user in order to apply the transmission strategy effectively. Further, LU is assumed to take action based on only local information, specifically the transmission state of the opponent in the previous time interval, during the cognitive transmission.

3.9.2 CONCLUSIONS

This chapter presents a game-theoretic approach for physical-layer security in large-scale IoT environments. We present an uncoordinated transmission strategy for IoT devices under jamming interference to meet information availability objectives while preserving limited transmission resources.

Utilizing a zero-determinant strategy, the limited-resource IoT devices are not required to know the transmission history of the adversary in order to apply their own transmission strategy; the strategy only requires knowledge of the most recent transmission actions of the adversary, which enables a decentralized scheduling scheme and reduces the coordination overhead for the IoT devices in large-scale deployments. Numerical results demonstrate the benefits of the proposed approach where the IoT devices can achieve their target information availability over time when adopting this transmission strategy.

REFERENCES

1. Ashraf Al Daoud, George Kesidis, and Jorg Liebeherr. Zero-determinant strategies: A game-theoretic approach for sharing licensed spectrum bands. *IEEE Journal on Selected Areas in Communications*, 32(11):2297–2308, 2014.
2. Manos Antonakakis, Tim April, Michael Bailey, Matt Bernhard, Elie Bursztein, Jaime Cochran, Zakir Durumeric, et al. Understanding the Mirai Botnet. In *USENIX Security Symposium*, Vancouver, BC, Canada, pp. 1093–1110, 2017.
3. Ali Ismail Awad and Jemal Abawajy, editors. *Security and Privacy in the Internet of Things: Architectures, Techniques, and Applications*. John Wiley & Sons, Hoboken, NJ, 2021.
4. Ribhu Chopra, Chandra R. Murthy, and Ramesh Annavajjala. Physical layer security in wireless sensor networks using distributed co-phasing. *IEEE Transactions on Information Forensics and Security*, 14(10):2662–2675, 2019.
5. Ashutosh Dutta and Eman Hammad. 5G Security challenges and opportunities: A system approach. In *IEEE 3rd 5G World Forum (5GWF)*, Bangalore, India, pp. 109–114, 2020.

6. Abdallah Farraj. Impact of cognitive communications on the performance of the primary users. *Wireless Personal Communications*, 71(2):975–985, 2013.
7. Abdallah Farraj. Analysis of primary users' queueing behavior in a spectrum-sharing cognitive environment. *Wireless Personal Communications*, 75(2):1283–1293, 2014.
8. Abdallah Farraj. Switched-diversity approach for cognitive scheduling. *Wireless Personal Communications*, 74(2):933–952, 2014.
9. Abdallah Farraj. Cooperative transmission strategy for industrial IoT against interference attacks. In *IEEE Texas Power and Energy Conference (TPEC)*, College Station, TX, USA, pp. 1–6, 2023.
10. Abdallah Farraj. Coordinated security measures for industrial IoT against eavesdropping. In *IEEE Texas Power and Energy Conference (TPEC)*, College Station, TX, USA, pp. 1–5, 2023.
11. Abdallah Farraj, Eman Hammad, Ashraf Al Daoud, and Deepa Kundur. A game-theoretic control approach to mitigate cyber switching attacks in smart grid systems. In *IEEE International Conference on Smart Grid Communications (SmartGridComm)*, Venice, Italy, pp. 958–963, November 2014.
12. Abdallah Farraj, Eman Hammad, Ashraf Al Daoud, and Deepa Kundur. A game-theoretic analysis of cyber switching attacks and mitigation in smart grid systems. *IEEE Transactions on Smart Grid*, 7(4):1846–1855, 2015.
13. Abdallah Farraj and Scott Miller. Scheduling in a spectrum-sharing cognitive environment under outage probability constraint. *Wireless Personal Communications*, 70(2):785–805, 2013.
14. Ivan Farris, Tarik Taleb, Yacine Khettab, and Jaeseung Song. A survey on emerging SDN and NFV security mechanisms for IoT systems. *IEEE Communications Surveys & Tutorials*, 21(1):812–837, 2018.
15. Kamal Gulati, Raja Sarath Kumar Boddu, Dhiraj Kapila, Sunil L. Bangare, Neeraj Chandnani, and G. Saravanan. A review paper on wireless sensor network techniques in Internet of Things (IoT). *Materials Today: Proceedings*, May 2021.
16. Mardiana binti Mohamad Noor and Wan Haslina Hassan. Current research on internet of things (IoT) security: A survey. *Computer Networks*, 148:283–294, 2019.
17. Line Larrivaud. State of enterprise IoT security in North America: Unmanaged and unsecured. A Forrester Consulting Thought Leadership Paper Commissioned By Armis Inc, September 2019.
18. Yi Luo, Ferenc Szidarovszky, Youssif Al-Nashif, and Salim Hariri. Game theory based network security. *Journal of Information Security*, 1(01):41, 2010.
19. Chris Y. T. Ma, Nageswara S. V. Rao, and David K. Y. Yau. A game theoretic study of attack and defense in cyber-physical systems. In *IEEE Conference on Computer Communications Workshops (INFOCOM WKSHPS)*, Shanghai, China, pp. 708–713, April 2011.
20. Moustafa Mamdouh, Ali Ismail Awad, Ashraf A. M. Khalaf, and Hesham F. A. Hamed. Authentication and identity management of IoHT devices: Achievements, challenges, and future directions. *Computers & Security*, 111:102491, 2021.
21. Fatma Masmoudi, Zakaria Maamar, Mohamed Sellami, Ali Ismail Awad, and Vanilson Burégio. A guiding framework for vetting the internet of things. *Journal of Information Security and Applications*, 55:102644, 2020.
22. Amitav Mukherjee. Physical-layer security in the internet of things: Sensing and communication confidentiality under resource constraints. *Proceedings of the IEEE*, 103(10):1747–1761, 2015.

23. Nahla Nurelmadina, Mohammad Kamrul Hasan, Imran Memon, Rashid A. Saeed, Khairul Akram Zainol Ariffin, Elmustafa Sayed Ali, Rania A. Mokhtar, Shayla Islam, Eklas Hossain, and Md Arif Hassan. A systematic review on cognitive radio in low power wide area network for industrial IoT applications. *Sustainability*, 13(1):338, 2021.
24. Stephen S. Oyewobi, Karim Djouani, and Anish Matthew Kurien. A review of industrial wireless communications, challenges, and solutions: A cognitive radio approach. *Transactions on Emerging Telecommunications Technologies*, 31(9):e4055, 2020.
25. Harold Vincent Poor and Rafael F. Schaefer. Wireless physical layer security. *Proceedings of the National Academy of Sciences*, 114(1):19–26, 2017.
26. William H. Press and Freeman J. Dyson. Iterated Prisoner's Dilemma contains strategies that dominate any evolutionary opponent. *Proceedings of the National Academy of Sciences*, 109(26):10409–10413, 2012.
27. Zhengguo Sheng, Shusen Yang, Yifan Yu, Athanasios V. Vasilakos, Julie A. McCann, and Kin K. Leung. A survey on the ietf protocol suite for the internet of things: Standards, challenges, and opportunities. *IEEE Wireless Communications*, 20(6):91–98, 2013.
28. Ali Hassan Sodhro, Ali Ismail Awad, Jaap van de Beek, and George Nikolakopoulos. Intelligent authentication of 5G healthcare devices: A survey. *Internet of Things*, pp. 100610, 2022.
29. Statista Research Department. Internet of Things (IoT) connected devices installed base worldwide from 2015 to 2025, 2016.
30. Lu Sun, Liangtian Wan, Kaihui Liu, and Xianpeng Wang. Cooperative-evolution-based WPT resource allocation for large-scale cognitive industrial IoT. *IEEE Transactions on Industrial Informatics*, 16(8):5401–5411, 2019.
31. Pal Varga, Jozsef Peto, Attila Franko, David Balla, David Haja, Ferenc Janky, Gabor Soos, Daniel Ficzere, Markosz Maliosz, and Laszlo Toka. 5G support for Industrial IoT Applications: Challenges, solutions, and research gaps. *Sensors*, 20(3):828, 2020.
32. Zhongxiang Wei, Christos Masouros, Fan Liu, Symeon Chatzinotas, and Bjorn Ottersten. Energy-and cost-efficient physical layer security in the era of IoT: the role of interference. *IEEE Communications Magazine*, 58(4):81–87, 2020.
33. Quanyan Zhu and Tamer Basar. A dynamic game-theoretic approach to resilient control system sesign for cascading failures. In *International conference on High Confidence Networked Systems*, pp. 41–46, April 2012.

4 Policy-Driven Security Architecture for Internet of Things (IoT) Infrastructure

Kallol Krishna Karmakar, Vijay Varadharajan, and Uday Tupakula
University of Newcastle

4.1 INTRODUCTION

The Internet of Things (IoT) is being touted as the next big technological trend by both academic researchers and industry players. The IoT refers to the connection of 'things' in the real world via the internet so that these 'things' can communicate with each other as well as other internet services. Thus, IoT enables the connection of the physical world with the cyber world, allowing remote monitoring and control of physical objects from the cyber world [15,19].

The IoT is increasingly being used in different applications ranging from precision agriculture to critical national infrastructure by deploying many resource-constrained devices in often unmanned and untrusted environments [41]. Such devices are becoming prevalent due to the ability to integrate the data from these devices into applications leading to significant benefits such as in advanced manufacturing, smart homes, and smart infrastructures [7]. In addition, the integration of physical actuators controlling the real world enables IoT applications to control smart environments. In general, smart environments are dynamic in that new devices are added and old devices are removed when they become obsolete or break down or are moved from one location to another. Also, there is usually a corresponding digital representation of these devices, which needs to be synchronised with the physical devices for reliable monitoring. For this to occur, devices and actuators need to be registered and connected/bound to IoT middleware, so that data can be extracted from these devices and processed by the IoT applications, as well as to receive control commands from IoT applications. This process is referred to as the *provisioning* of IoT devices [1,29]. IoT middleware is a software/script/API or software glue that interfaces IoT device components and system applications and enables communication between them. It helps to resolve the issues of physical-layer communications, application service requirements, and diversity/heterogeneity in communications [4].

 DOI: 10.1201/9781003199410-4

The drawback of the device onboarding process is that it may cause substantial risk to the IoT network infrastructure. During onboarding, a device exchanges its capabilities (e.g., cipher-suite and handshake protocol) with the connecting network infrastructure using a hub or a gateway or a network controller. Often, default credentials are hard coded and exchanged during communications. An adversary can intercept and weaponise these communications for attacking the IoT infrastructure or the devices. For instance, in a smart agricultural farm, monitoring devices can be de-authenticated due to IEEE 802.11 vulnerabilities [37]. Furthermore, the resource-constrained nature of the IoT devices makes them susceptible to distributed denial of service (DDoS) attacks (e.g., continuous and multi-threaded device capability requests consuming the device buffer and processing resources, putting the device into a stalled state). In some cases, an adversary can launch severe attacks by compromising vulnerabilities/flaws in the network infrastructure and devices.

An advanced persistent threat (APT) launched against industrial IoT infrastructure is a classic example of such an attack [39]. Such attacks can be hard to detect due to the lack of attack signatures and poor security posture of the low budget smart infrastructures. Moreover, the devices that are positioned in remote locations can be compromised (physically) during runtime. For instance, consider a weather sensor-grid monitoring changes in the coastal power plants. An adversary can compromise the sensor grid and send fake weather warnings to the power plants. Hence, there can be a need for real-time monitoring for smart devices and sensors. The infected devices can also act as bots sending malicious packets and communication requests flooding the network infrastructure [2]. In addition, due to the resource-constrained nature of IoT devices, an attacker can target IoT devices easily by sending forged requests, intercepting and illegally manipulating valuable sensor data in transit, capturing a physical device, and transforming it into a zombie to launch attacks on other systems within the network infrastructure. Denial of service (DoS) and energy depletion attacks are the most common IoT attacks [8,32]. Often, it is not possible to implement security on these devices using traditional defence mechanisms as they are located in open environments, and they would incur extra computational load on small IoT devices. Hence, a secure IoT network infrastructure with secure device provision is a requirement.

Secure provisioning of IoT devices enables the digital representation of the devices to be synchronised with the physical devices and is a key design issue in the development of smart IoT-based environments, as security risks to the digital representations can lead to harmful impacts on the physical environment [22]. Also, a secure programmable IoT network infrastructure is necessary for future networks. Therefore, a secure smart IoT environment requires secure provision and automated IoT devices' onboarding. Furthermore, the secure IoT device provisioning and management require security policies and mechanisms to control and specify the malicious or suspicious device activities within the smart network infrastructure.

In recent times, elliptic curve cryptography (ECC)-based device authentication and provisioning protocols are becoming popular [22,44,46]. In some cases, protocols are modified to suit the environment and device heterogeneity [46]. However, traditional authentication and authorisation mechanisms fail to locate the malformed devices with dormant malware. Also, after authenticated provisioning, malicious activities of such devices still remain a major concern. Thus, we need mechanisms not only for secure provisioning but also for controlling the malicious behaviour following device provisioning. Our consolidated IoT security architecture can also ensure the security of IoT network infrastructure assets.

In this chapter, we describe our approach to secure provisioning/monitoring and policy-aware secure management of IoT devices. We propose a consolidated security architecture, secure device provisioning, and secure IoT network environment management. The proposed security modules and services can be used to deploy IoT provisioning and management using cloud infrastructure. Our approach has the following distinct features, making it suitable for different practical applications.

Our architecture presents a policy-driven approach to secure device provisioning and secure IoT network infrastructure management. Firstly, it specifies pre- and post-condition policies that relate to the attributes and state of devices, which are used to enforce specific security constraints while provisioning devices. They are also used in extracting and providing sensor data to IoT applications and receiving control commands from them. This also enables us to specify what type of scripts can run on a device as well as the authorised communications that a device can have. Secondly, our architecture incorporates the notion of digital twin synchronising with the physical device, which is used to monitor and manage the security state and health of the devices using cloud-based applications. Thirdly, the proposed security architecture offers fine granular policies to manage the flow communication of the IoT network infrastructure. Finally, our security architecture can ensure on-demand security services such as confidentiality to the device flows.

To demonstrate the efficacy of the proposed security architecture, we present a smart agricultural farm infrastructure scenario. In principle, the proposed approach applies to the deployment and management of IoT devices and associated network infrastructure in various applications such as smart homes, healthcare environments, and industrial IoT environments.

We have structured the chapter into the following sections. Section 4.2 discusses relevant related works and compares them with our approach and security architecture. Section 4.3 introduces fundamentals and benefits of policy-based network management. Section 4.4 presents an IoT network scenario and its associated challenges. This section also explains the device ontology, which describes the attributes, properties, and binding information associated with the sensor and actuator components of a device. Section 4.5 describes our policy-driven security architecture approach. This section presents two major modules and associated components of the security architecture. This section outlines the pre- and post-conditions associated with the secure device provisioning process. Also, it introduces fine granular policies for IoT network infrastructure management. Section 4.6 describes the prototype implementation and discusses how it counteracts the different attack scenarios. It also presents performance evaluation, security analysis, and comparison. Section 4.7 presents an open discussion on this work and some possible future extensions of this work. Finally, Section 4.8 concludes this chapter

4.2 RELATED WORK

We have classified the related works into two sections: first, we will discuss automatic device provisioning architectures and then about security services associated with the provisioning mechanism.

4.2.1 POLICIES AND SDN

In RFC 1102, Clark introduced policy-based routing for autonomous domains [11], which proposed a simple policy syntax for interdomain communications. We have refined and extended the policy syntax to develop fine-grained security policy specifications targeted for IoT devices and SDN network characteristics.

Das et al. [13] present a context-sensitive policy framework for IoT devices, to control and protect information sharing between them. Their policies capture the diverse nature of IoT devices and their interaction with network users using an attribute-based access control policy. Their work mostly focuses on the privacy of user data. In our case, we have focused on securing the IoT network infrastructure using fine granular access policies. Beetle [23] is an access control policy framework for operating systems (Linux, Android) to control application interaction with peripheral device resources and provides transparent access to network devices. Later, Hong used the Beetle framework in home network gateways to control IoT communications [18]. This work does not address the authentication of IoT devices, or users, which our architecture does consider. Other work on access control policies for IoT devices can be found in Refs. [24,30,45]. However, none of these works have used SDN to manage and enforce the policies. Furthermore, they are mainly concerned with user security rather than IoT security.

4.2.2 AUTOMATIC DEVICE PROVISIONING

Hilmer et al. [17] presented a device provisioning architecture and modelling approach for smart environments with their sensors, actuators, and devices. This work focuses on the modelling and assuring easy deployment and connectivity of smart sensor devices. Our work extends the proposed architecture and adds security services for the secure provisioning of these smart sensor devices. Such security services enable SDPM to defend against different attacks towards smart sensor devices and infrastructure.

A multipurpose binding and provisioning platform (MBP) was presented by Silva [12]. The platform provides an easy open-source solution to provision the interconnected devices and manages them efficiently using apps. LEONORE is a service-oriented elastic provisioning infrastructure for large-scale IoT devices [43]. This provisioning architecture focuses on provisioning the application components associated with the devices. Provisioning the industrial IoT devices manually is very troublesome due to its environmental complexity and huge numbers. Hence, Wang et al. [44] have presented a state machine-based device provisioning process approach that requires zero human interaction. They are using the remote authentication dial-in user service (RADIUS) protocol and a one-time password to authenticate the devices. The above-mentioned approaches consider only the ways to auto provision the devices. However, as IoT devices are becoming more available, the attack surfaces for such devices are increasing. Our SDPM architecture considers the IoT devices state (internal/external) and environmental contexts to provision the devices. We have introduced granular device profile attributes, which help us to create precision policies to limit IoT operation. The pre-/post-condition policies utilise such granular attributes to check the devices and its communication for security purposes. Such an approach allows us to defend the IoT devices and IoT network infrastructure from certain attacks.

P. Zhang et al. [49] has proposed a way to detect faulty nodes in a fog system using the state transition. They are using continuous Markov chain modelling in state transitions to detect faulty nodes. Our work is entirely different from theirs, as we are using a whitelist-based approach to detect the device state change. C. Zhang et al. [48] presented how EEG can be used to assess the driver's state while they are driving. The study uses a learning-based algorithm to accurately train and detect the driver state. It focuses on capturing biological signals (EEG) to determine the state/condition of a human driver. Our work focuses on assessing the state of the OpenFlow switch. We define the state of the device in terms of the internal and external processes running in them and the condition of the I/O communication buffer. Hence, we proposed a security architecture that collects such information and uses it to make further decisions. The current work does not focus on using learning algorithms; instead, it uses pre- and post-condition policies for its actions.

4.2.3 SECURE DEVICE PROVISIONING

Kohnhauser et al. [20] presented a secure provisioning service for industrial devices. Their approach focuses on open platform communications unified architecture (OPC UA) and utilises device certification to provision them securely. Our work uses more than device certificates. We utilise device and network context-specific pre- and post-condition policies that allow secure provisioning of the devices. The device context extends the authentication services with device state policies that can assess a device's trust status before provisioning it. This allows the network infrastructure manager to find malicious devices and defend the network from future attacks.

Software defined provisioning (SDP) is a software-defined network (SDN)-based IoT device provisioning architecture that provides scalability for IoT deployments [26]. This architecture also ensures robust dynamic authorisation and provisioning of heterogeneous IoT devices. Our architecture does not utilise the SDN. However, it can be integrated with SDN with some minor modifications. But the major strength of our work is granular pre- and post-condition policies. Furthermore, the policies are integrated with the digital twin to check the security status of the devices. Hence, our architecture is able to provide robust authentication and authorisation.

Sousa et al. [38] presented a provisioning architecture for IoT devices that uses certification authority and a one-time password to provision the devices. Their main focus is to tackle scalability issues. As mentioned previously, our provisioning services use device certification as one of the ways to test the legitimacy of the devices.

4.2.4 MACHINE LEARNING-BASED CLASSIFICATION OF DEVICES

There are many papers that address identification and classification of IoT devices using machine learning. With device provisioning in smart infrastructures, we require authentication (to ensure that the correct device is properly identified and authenticated) and authorisation (to ensure the device is behaving properly and its actions are legitimate). In studies such as [28], fingerprints are used as identifiers, and machine learning is used to classify these communications and devices as benign and malicious. For instance, in Ref. [3], LSTM-CNN cascade model is used to classify the IoT device communications into different categories (such as hub, electronics, and cameras). Note that such works emphasise device identification based on their communication and behavioural characteristics. Our work is different to these works as it is about authentication of devices based on their attributes and state and their verification using secure credential and certificates, and then using policies for their provisioning and their subsequent management.

4.2.5 IoT SECURITY AND ATTACKS

Pongle et al. [33], Lyu et al. [25], and Mendoza et al. [27] consider various attacks to compromise IoT networks. These works focus on analysing the vulnerabilities in IoT network infrastructure, whereas our work is mainly concerned with the design of authentication and authorisation policy-based security architecture for IoT network infrastructure. Pa et al. [31] provide an analysis of Telnet-based attacks on IoT devices. They propose an IoTPot (a honeypot) and IoTBox (a sandbox), which help to attract Telnet-based attacks against various IoT devices running on different CPU architectures. Their work mostly focuses on analysing the IoT malware threats. In this study, we have provided a detailed analysis of Mirai attack. Moreover, our security architecture helps to block telnet-based attacks and prevent their spreading. Capellupo et al. [9] present an analysis of present home automation devices such as Amazon Echo and TpLink smart plug and discuss how they can pose major security threats to home networks and to user privacy. Their work focuses mainly on addressing different threat vectors for the home IoT network infrastructure. Our approach helps to rectify some of the problems mentioned in their work; for instance, our architecture can help to prevent unauthorised users/devices gaining access to network services and IoT device traffic.

4.3 FUNDAMENTALS OF POLICY-BASED NETWORK AND SECURITY MANAGEMENT

The size of network domains is increasing day by day. Each domain consists of a huge number of users, resources in the form of heterogeneous devices (routers, wired and wireless gateways, IoTs, servers, etc.), applications (cloud management tool, monitoring tool, security services, QoS assurance tools, etc.), new networking technologies, and services (on demand multimedia request, online gaming, etc.). These users, resources, and services inside any network domain are dynamic, and they can be easily removed, added, or updated during runtime, thus incurring a huge amount of complexity when it comes to the automatic management and scheduling of network domain and services, for instance, a huge industrial network equipped with thousands of time-critical IoT equipment and users using them on demand wirelessly. Any delay in scheduling the operation of these IoT devices will cause serious damage to the industrial and production infrastructure. Thus, an automated approach is needed to control the scheduling as prescribed earlier. This scheduling problem is only one end of the spectrum of different network domain problems due to the massive number of users requesting various services through different hardware configurations. Reduced bandwidth, low QoS, and delays in multimedia are the most common ones. To deal with these issues, policy-based management for network infrastructures was introduced back in 1999 with RFC2026 [10,21,35,40].

4.3.1 POLICY

A policy is a set of rules that consists of some criteria or conditions with appropriate actions defining the behaviour of a system, if the particular criteria are successful. The criteria are declarative and depend on the nature and availability of resources of that particular system. The actions are tasks that need to be enforced or administered by the system. Each policy defines how various resources within a system can be accessed and used if a set of criteria are fulfilled.

4.3.2 POLICY-BASED NETWORK AND SECURITY MANAGEMENT

It is a paradigm where policies are used to control and manage the security and resources of network infrastructure. It facilitates flexibility in the maintenance of large and complicated network infrastructure with less effort. It helps to control the dynamic nature of future network infrastructures. It is easily deployable, requires less maintenance, is cost-effective, provides better performance, and is adaptable during runtime, and hence, it is becoming highly popular.

Now, we will describe the common architecture for the policy-based system and its benefit.

4.3.3 POLICY-BASED MANAGEMENT ARCHITECTURE

IETF and DMTF have developed a policy core information model (PCIM) (Figure 4.1). It has four major components: (i) Policy Repository (PR), (ii) Policy Management Tool, (iii) Policy Decision Point (PDP), and (iv) Policy Enforcement Point (PEP).

4.3.3.1 Policy Repository (PR)

The storage/database where the policies are stored is known as the policy repository. It can vary from a simple file to a huge data storage server based on the context of the management environment. For instance, a social media platform like Facebook uses a huge warehouse as a policy storage server, whereas a small homes IoT device management uses one JSON file to store all the management policies in an MQTT server. Here, this single file acts as a policy repository (PR).

A policy repository serves various purposes in a policy-based management architecture, such as (i) storing the polices, (ii) querying the stored policies, (iii) retrieving of stored policies, (iv) resource validation (constraint checking), (v) policy translation, (vi) policy transformation, and (vii) securely storing the policies.

The policy repository has to be very fast. The performance of the whole policy-based management system in some cases depends on the speed of the policy repository. Currently, SSDs are used to store policies in large warehouses for quicker query and fast retrieval of them.

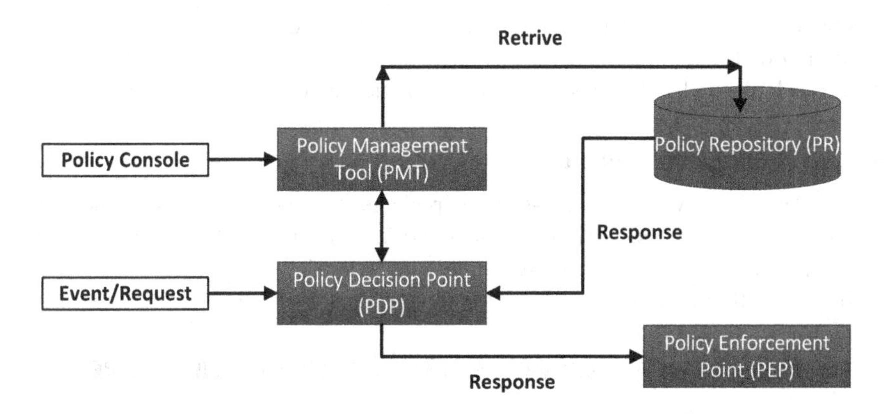

Figure 4.1 The figure presents a fundamental policy-based management architecture. It has four major components that help to retrieve information, store policies, formulate decision, and enforce the policy.

4.3.3.2 Policy Management Tool (PMT)

A server or host where the Policy Management software resides is known as the Policy Management Tool. It helps in providing an interactive and user-friendly graphical or web user interface to easily edit and represent the stored policy. In some cases, the PMT can accept input from the policy consoles as well. The policy console is a terminal like interface to send CRUD instructions to the PMT.

The PMT maintains the whole policy ecosystem. It helps in the retrieval of the appropriate policy from the PR. The PMT uses Lightweight Directory Access Protocol (LDAP) to maintain the communication with the PR [34]. LDAP is a vendor-neutral, open source application protocol to maintain access and manage information resources over an IP network. This protocol provides all the features and instructions of a database management environment such as add, delete, update, search, and bind. The PMT also helps in the translation and validation of the policies.

One of the major problems with the policy-centric ecosystem is that they are victim to conflicts. For instance, in a home thermostat, a policy says that if the temperature goes higher than 20 degrees turn it on. On the other hand, for that same home thermostat, another policy says, it should be turned off. The PMT helps to resolve such policy conflicts raising alerts to the user interfaces during the installation or while executing one of the policies.

4.3.3.3 Policy Decision Point (PDP)

The PDP is the monitoring point for this architecture. The PDP evaluates the necessity of policy enforcement for any certain event. The PDP captures the events and requests within a system and consults with PMT to retrieve the matching policy from the PR. It is an intermediary that can translate the events/request readable by the PMT. It also helps to check whether the policies are enforced correctly or not.

4.3.3.4 Policy Enforcement Point (PEP)

The PEP is a point where the PDP-selected policies are to be enforced. For a network infrastructure, gateways, switches, routers and WAPs act as PEP. Sometimes, PDP and PEP can act as a single entity residing in the same device.

It helps with the validation of policies and captures feedback from the system.

4.3.4 BENEFITS OF A POLICY-BASED MANAGEMENT ARCHITECTURE

The following are the benefits of a policy-based management architecture:

Managing the complexity: A system is a collection of different entities, for instance, an enterprise network consists of smart devices, users, end-host machines, servers, gateways, and switches. Policy-based management architecture provides a way to classify entities in a system into different groups and can enforce management policies into these groups. We can consider the example of the enterprise network where it has two VPN networks: one is secure (encrypted channel) and the other one is normal. Based on the user's role, their communications are classified into two groups, namely, classified and unclassified. No matter which devices they log-in from, the network policies will provide the associated services to the respective users. Thus, policy-based management helps to simplify device, user, network, service management, or any distributed system complexities with less effort.

Less human effort: With policy-based management architecture, it requires less personnel to configure the network or system environment. Once the polices are installed, the system should function according to the installed policy. Thus, it requires less human resource and effort.

Time-critical functions: One of the most efficient ways to implement time-critical functions is policies. For instance, it can enforce timed network policies that can restrict user's network communication for a certain duration.

Better security: As the systems become big, it becomes more complex. In these complex systems, the chances of resource abuse are very high. For instance, in a networked system, malicious users can launch DDoS attack to disrupt the network operation. One of the best ways to resolve resource abuse is to use policies.

4.4 IoT NETWORK SCENARIO

This section presents an IoT network scenario and lays out the security policy requirements for such infrastructure.

Figure 4.2 presents an IoT network infrastructure. The IoT network is managed and controlled by a programmable network controller. There will be IoT sensors and actuators connected to this network. Each IoT sensor/actuator is connected to a local IoT node. Multiple IoT nodes are connected to IoT gateways. The IoT gateways are connected to routers or switches. Sometimes, gateways and switches are combined and can be either wired or wireless. Finally, all the IoT sensors/actuators upload the data to the cloud infrastructure.

Based on the IoT network operation, there are multiple locations where security policies can be applied to fulfill the following requirements:

- **User-specific policy**: Who can access which device under which condition?
- **Device-specific policy**: A. Which device can access what network service under which condition?
  B. What is the state of the device? C. Which device under what state can serve requests from particular users?
- **IoT/service route-specific policy**: A. Which device, owned/used by the user/-domain/service, can access various network services via which network path? B. What conditions do the routing devices need to satisfy?
- **Cloud service-specific policy**: A. What service can serve a particular IoT device and User under which condition? B. What should be the flow communication state?

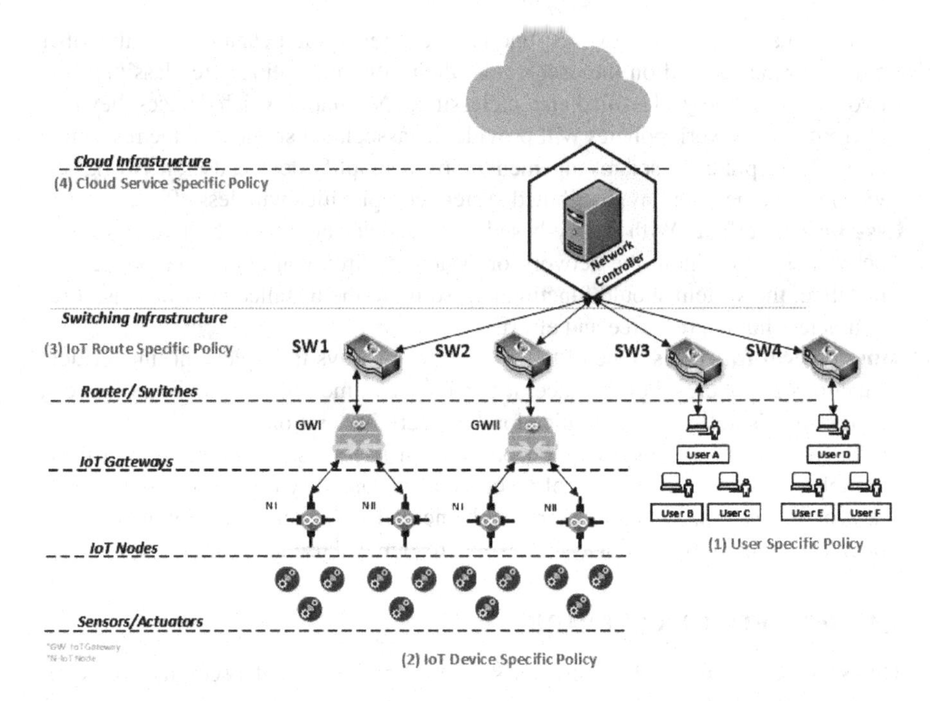

Figure 4.2 The figure presents an IoT scenario where multiple sensors/actuators are connected to a programmable network infrastructure. They upload data to the cloud infrastructure. Users can control these devices using cloud service-specific apps.

4.4.1 TYPES OF DEVICES AND DEVICE ONTOLOGY

4.4.1.1 Types of Devices

IoT devices can be diverse, depending on the type of protocol they use and its purpose of use. For instance, LoWAN, 6LoWPAN, and ZigBEE are different types of protocols supported by the IoT devices. IoT devices used in the medical domain, agriculture domain, and industry domain are of various types as they are purpose-built and targeted to achieve a specific goal. However, here we focus on the connectivity of the devices, i.e., how they are connected to the network infrastructure. Our approach for the provisioning of devices envisages the devices to be either plug-and-play or configurable. A plug-and-play device has embedded sensors and actuators and provides interfaces to access sensor data and control it through its actuators. Configuration of this type of device is not possible. Examples of such devices are WiFi-enabled wearables, cameras, and audio systems.

A configurable device has sensors and actuators attached to it and offers a runtime, e.g., to deploy device adapters that extract and provision sensor data. An example of such a device is a Raspberry Pi. There can also be constrained configurable devices with limited processing and storage capabilities, which can be connected to more powerful systems such as a Gateway. Arduino is an example of such a constrained device. To classify the IoT devices more precisely, we introduce a device ontology now.

4.4.1.2 Device Ontology

A device ontology is a meta-data based descriptive representation of a device, its capabilities, and its properties. The device ontology describes the details necessary for device registration, binding, and enabling access to the devices; it can also be used as a meta-data source by IoT applications. A device consists of a sensor, actuator, adaptor, and its state. The device ontology contains meta-data, including sensor and actuator specifications such as their accuracy and frequency and many more. The use of ontologies in IoT is widespread due to the heterogeneous, dynamic environments that have to be integrated. The device-related information in ontology allows the security services running in the network infrastructure to perform preliminary security assessments while on-boarding the device. The device ontologies are extensible, which can take into account new processes and features.

In our current security architecture, device ontology will help formulate the pre- and post-conditions required for the policy-based approach to secure IoT devices provisioning.

Now, we present how the ontology can help to present a device. Figure 4.3 presents a schematic representation of such an ontology.

Sensor and actuator: A device can be equipped with both sensors and actuators. Each sensor/actuator consists of a name and an identifier. There can also be other attributes associated with a device and its sensor/actuator module. Their name, ID, types, location of use, allocated channel, etc., are most common attributes a sensor/actuator has. Apart from that, a device manufacturer can include sensor/actuator quality-specific attributes. For instance, a water level indicator sensor can have quality-specific attributes such as accuracy and sensitivity. There can be certificates associated with a device from the manufacturer, which will be useful for device attestation during boot time from a security perspective.

State: Each device has an operating system or firmware that controls its function or behaviour. Input to the sensor or other internal and external events are linked with the IoT device's state. The state of a device will also include the state of its internal soft/hard components, i.e., OS, physical memory, and IO state. Hence, from a security and trust perspective, a device is subjected to external and internal events. For instance, an external event could request to read data from its sensor or perform an actuator's action. The internal events are related to the running of its operating system, for instance, an I/O driver handling the input and output operation of a Raspberry Pi device using Raspbian OS.

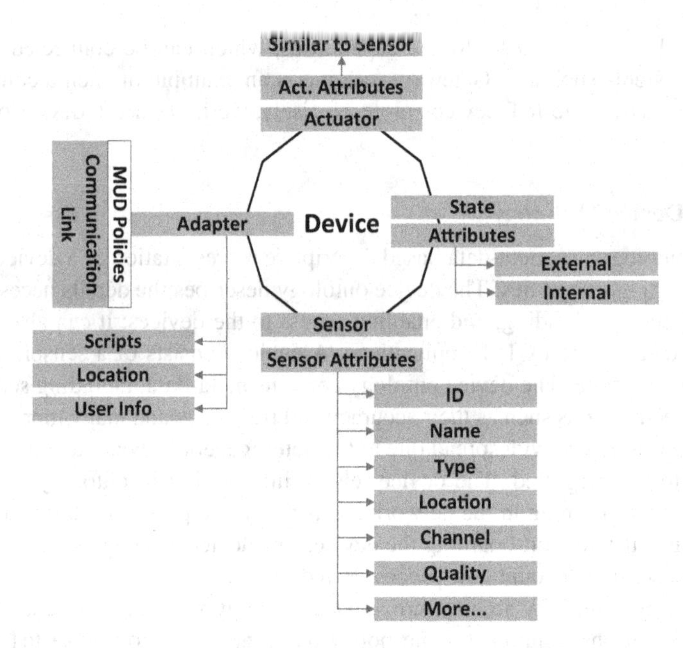

Figure 4.3 Device ontology is a meta-data based descriptive representation of a device, its capabilities, and its properties. This figure presents the device ontology for the secure provisioning architecture. A device can be a sensor or actuator. Each will have associated attributes. The adapter attributes present communication options, and the state attributes present the device status.

Adapter: An adapter is a communication socket that the devices use to communicate with the network infrastructure. This is a set of libraries that helps the devices to understand the communication protocol, build, transmit, and receive data packets to and from the devices [6,17,29]. We use the adapter to capture MUD policies, which focus on communications of devices. For instance, a MUD policy for a smart bulb specifies that it should not communicate with the smart meter or the smart refrigerator. Instead, it should communicate with the update server of the manufacturer. These are specified in the form of access control lists for communications.

However, MUD policies do not address device state-based authorisations used to specify policies such as those needed for addressing patch vulnerabilities for devices or enforcing trust-related policies based on the state of the device. Our approach is able to specify not only MUD-based policies (as shown in Figure 4.3) but also more fine-grained policies based on attributes and state of the device as well as those involving attributes of the network environment. This helps us to specify policies capturing the requirements of different IoT applications as well as those needed in secure provisioning and management of devices.

4.5 POLICY-DRIVEN SECURITY ARCHITECTURE

Now we will present the policy-driven security architecture for IoT infrastructure. The security architecture focuses on solving the following challenges (Figure 4.4).

- **Challenge 1**: Security provision the IoT devices.
- **Challenge 2**: Manage resource-constrained IoT devices securely in an IoT network infrastructure.

4.5.1 DEVICE PROVISIONING?

The process of on-boarding an IoT device into a network infrastructure is termed as provisioning. The simplest form of provisioning is just an IP assignment, enabling the device to communicate with the rest of the network infrastructure devices and

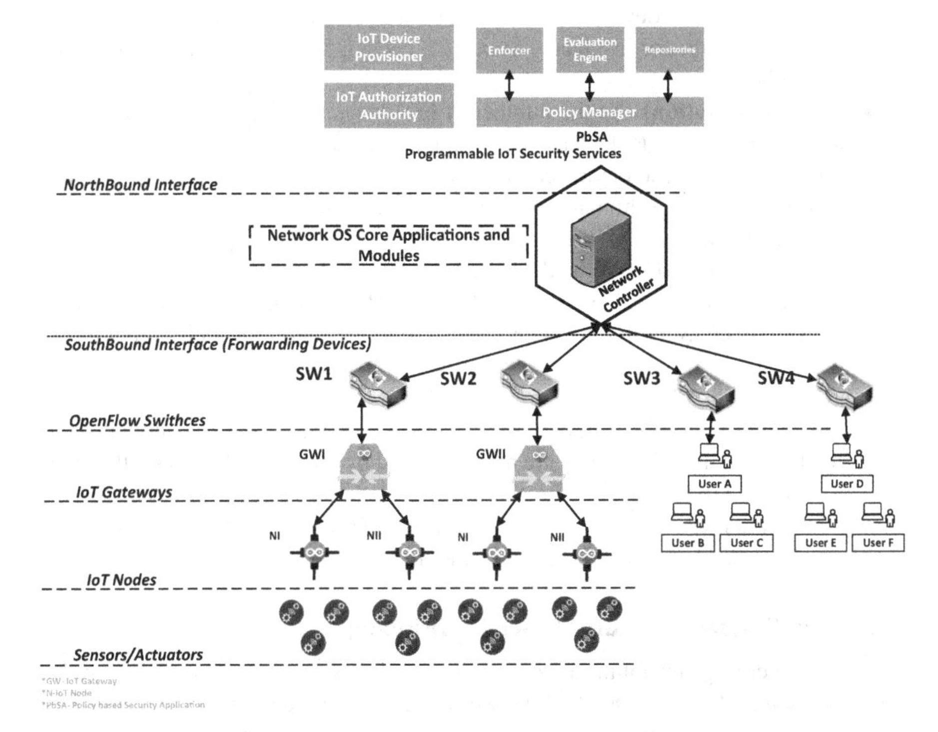

Figure 4.4 The picture presents a programmable network architecture and associated security services to secure the IoT network infrastructure from network attacks. The security service, as well as the network infrastructure, is being represented at an abstract level. Hence, it can be customised to suite the infrastructure requirements.

cloud. However, such simple provisioning may lead to potential security issues. A classic example is malicious devices being provisioned in the network. Another example is resource-constrained devices leaking provisioning credentials that can be used by the adversary to launch further attacks on the network. In this section, first, we will present the need for and requirements for secure device provisioning and then explain what policy-based device provisioning is.

Requirements of secure device provisioning include the following:

- **R1 (Identity)**: Nowadays, devices or their firmware can be physically tampered before they are being used in any smart network infrastructure. Once these rogue devices are provisioned into any smart network infrastructure, they will act as a gateway for any adversary to launch further attacks on the network domain. Hence, we need a proper mechanism to verify the identity of the device.
- **R2 (Authentication)**: Lack of proper authentication will allow the adversary to inject fake IoT devices into any network infrastructure. This can lead to a potential hazard, for instance, "Stuxnet ", causing massive damage to nuclear powerplants.
- **R3 (Authorisation)**: As the number of IoT devices are increasing, managing and identifying rogue devices and their activities within the network is becoming critical. A rogue device can act like a zombie and can compromise other devices or intercept human communication and actions. Hence, the network must have a device flow authorisation mechanism that only permits legitimate flow actions to and from the devices.
- **R4 (Trustworthiness)**: Sometimes IDS and IPS systems fail to detect a malicious device. Such failures can occur due to the nature of their activities or the frequency in which they do such events. For instance, a smart speaker is equipped with a MIC, which continuously records the conversation and transmits them (randomly or in precise time) to an adversary cloud. Although such incidents are malicious, they are not monitored by the IDS/IPS, and so, they are not reported to the owner. The examples demonstrate that the devices can become malicious or untrustworthy to carry on certain privileged activities over time. Hence, there is a need to know the trust status of the IoT devices.

4.5.1.1 Policy-Based Secure Device Provisioning

To achieve secure provisioning of a device, we propose a policy-based approach. It is a mechanism/services in which network, security, and device attribute-specific policies are used to securely provision the devices into a smart network infrastructure. The policy specifies pre-conditions that need to be satisfied *before* a device is provisioned, whereas the post-conditions specify the set of conditions that the device needs to satisfy *after* the device's provisioning. The conditions in our policy based approach consider three aspects of a device, namely: (i) attributes of a device; (ii) properties based on the state of a device; and (iii) rules associated with a device. Such a granular policy-based approach enables us to capture a range of security constraints related to the secure provisioning of various IoT devices. We outline a security architecture to realise such a policy-based approach for secure provisioning of IoT devices. We envisage that such an architecture can be integrated with the existing mechanisms and infrastructure in a cloud platform (such as Azure IoT or AWS IoT Services), to build a proof-of-concept demonstrating secure provisioning of devices in a smart agriculture farm scenario.

We will now present the syntax and attribute granularity of the policy-based approach to secure provisioning of a device.

4.5.1.2 Security Policy Language for Provisioning Devices

- **Pre-conditions**: These are a set of conditions that a device needs to satisfy *before* being connected to the network.
- **Post-conditions**: These are a set of conditions that a device needs to satisfy *subsequent* to provisioning.

4.5.1.2.1 Pre-condition

Pre-conditions are based on attributes and properties associated with the device and attributes of users who can provision the device. A diagrammatic representation of pre-conditions is shown in (Figure 4.5)

- **User attributes**: A pre-condition can specify conditions on the users who can provision a device. A device can only be provisioned by users who satisfy these conditions. This can be specified explicitly by naming the users or groups or roles associated with a user or, more generally, in the form of set of attributes that the users must have. From a security point of view, these attributes need to be reliable such as user certificate (e.g., X.509 certificate) and secure credentials. This can be specified using role-based and attribute-based policies.
- **Device attributes**: A pre-condition can specify conditions on the attributes that a device must have before it can be provisioned. The device attributes include device type, device certificate, and certificate of the device manufacturer.
- **Device state properties**: A pre-condition can also specify conditions that the state of a device needs to satisfy. We will express this in the form of state properties. We will elaborate these properties further in the next section.

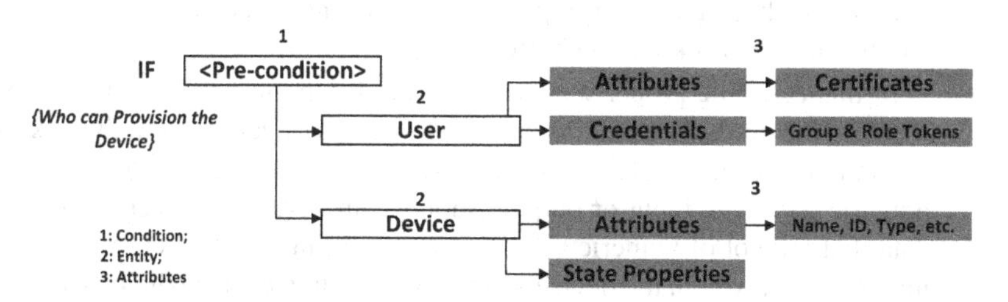

Figure 4.5 A visual representation of the pre-condition for the SDPM architecture. Here, 1 presents the conditions over entities and attributes represented by 2 and 3, respectively.

4.5.1.2.1.1 Pre-condition Examples In this section, we provide some examples of pre-conditions in the policies used to provision IoT devices.

We will denote a device as x and a valid user as u.

- **Using security attributes**: The device x can be provisioned if and only if the user is a valid user (u_i^x) and belongs to a valid group ($Group$), and the device is manufactured by a valid manufacturer. That is,
  - Device x can be provisioned by user u if u is a valid user *and* $type(x)$ is valid *and* $manufacturer(x)$ is valid

  For instance, u can be valid, if $cert(u)$ is valid and/or $u \in Group$ is valid. Similarly, $manufacturer(x)$ is valid if $cert(manufacturer(x))$ is valid.
- **Using state properties**:
  - **Property 1 - New device**: The device can be provisioned if it is a new device. This can be checked by examining the device log. If there is no device log (and device has never been used), it can be inferred that the device is new. Assume that this is reflected in a state variable *new*. Using this property, the condition can be expressed as follows:

    if ($new(x)$ is valid), *then* the device can be provisioned.
  - **Property 2 - Device internal process state**: The device ontology stores a set of benign and default processes for each IoT device in its database. During device onboarding, the SDPM uses device whitelisting with the help of the device ontology database to check the processes running in the device. Now,assume there is a reference set of processes available for a device (e.g., whitelisted processes for a device). Any new device that generates these whitelisted processes when it runs can be provisioned. That is,

    if ($\forall process(x) \in Reference\_Set$), *then* provision the device.
  - **Property 3 - Device external process state**: External communications of a device are reflected in the I/O buffer state. Applications and services running inside an IoT device communicate using network sockets/ports. These services continuously disseminate information (data/control) using these ports (essentially, it means writing information into the I/O buffer). Our security architecture considers such communication port behaviour as an I/O buffer state change. Network enumeration or probing tools (such as nmap) are used to check the network port status. Our implementation uses probing and fuzzing to check the I/O buffer state.

    In this case, the property is based on the I/O buffer state used to detect whether the device is in an insecure state (e.g., that can potentially leak information). For instance, assume a sensor's buffer is supposed to have only numeric data type input or output. Suppose an adversary injects malicious data that are not of numeric value (such as a string). In that case, this mismatch in data type in the device's buffer will be treated as malicious and the device will not be provisioned. That is,

    if ($i/o\_state(x)$ is valid), *then* provision the device.

4.5.1.2.2 Post-condition

Post-condition policies lead to actions that should be performed after a device is provisioned.

After the devices are provisioned, they are registered in the Provisioned Devices Database. The set of post-conditions and actions will be assigned to the device that has been provisioned in this database. The post-conditions can use environment parameters such as location and time, as well as involve state properties of a device Figure 4.6.

- **Environment parameter - location**: Post-conditions can use the location parameter to specify where a device should be provisioned. This can also be mapped to network reference position such as the IP address or the location of the gateway to which the device should be connected to.
- **Environment parameter - time**: Post-conditions can use the time parameter to enforce when a device should be operational. For instance, this condition can be used to detect potential misbehaviour or malfunctioning of a device.
- **Device state properties**: As in the case of pre-conditions, post-conditions can use a device's internal and external state (such as process list and I/O buffers) to enforce actions that can be used to detect misbehaviour or malfunctioning of a device. For instance, the post-condition can specify when a water level sensor is provisioned, if the sensor measurement were to cross the allowed maximum threshold, then this will lead to a generation of system alert.

4.5.1.2.2.1 Post-condition Examples Let us assume that the device x is provisioned with a set of post-conditions.

- **Environment parameter - location**: A device is provisioned at certain specific locations as specified in the post-condition, $\text{location}(x) = \{\alpha, \beta\}$.
- **Environment parameter - time**: When a heat-lamp device in the farm is provisioned, the post-condition can ensure that it operates only when the temperature is below a certain threshold. For example, a time-specific post-condition can be of the form $\text{operational\_time}(x) = T$, where T is a time duration between $(0100 < T < 0700)$.

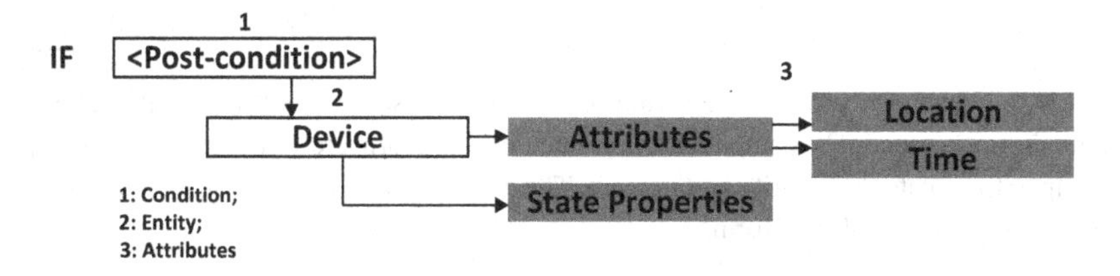

Figure 4.6 A visual representation of the post-condition for the SDPM architecture. Here, 1 presents the conditions over entities and attributes represented by 2 and 3, respectively.

- **Device state properties**: At the time of provisioning a water flow sensor device in a smart irrigation system, post-conditions can be used to detect (by recording the level in the logs) and prevent overflow (by generating an alarm when it passes the recommended water level). For instance, post-conditions can include:
  log_level if flowlevel(x) = Γ, where $2.5 < \Gamma < 3$ and *generate_alarm* if flowlevel(x) ≥ 3

4.5.2 SECURE SMART DEVICE PROVISIONING AND MONITORING SERVICE (SDPM)

Figure 4.7 gives an outline of the proposed components internal diagram. This internal diagram consists of a set of software components: Device provisioning gateway (DPG), provisioning evaluation engine (PEE), device provisioning service (DPS), policy database (PD), and provisioned devices database (PDD) with the post-conditions (POST-P). It has three modules and two databases to store the pre-condition policies and post-action/condition policies. Respective secure software modules can only access each database. Apart from that, we have introduced a digital twin of the device in the architecture. The digital twin's purpose is to represent the device status to the cloud and the user. Now we will present the functional description of each module in this architecture.

The software modules are the fundamental blocks of the security architecture, and the choice of implementing these modules depends on the type of smart network infrastructure and the connectivity to the cloud.

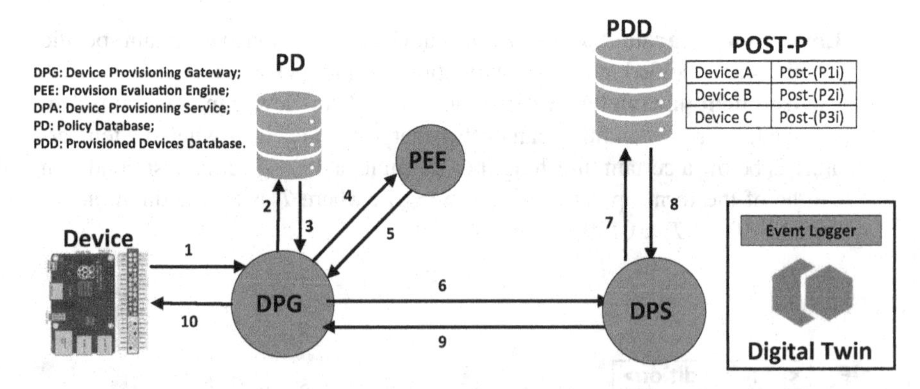

Figure 4.7 The figure presents the various components of secure smart device provisioning and monitoring service. The components presented in block form are software modules/scripts that help to formulate the security architecture. The devices can be physical or virtual, and they are represented in digital twin form.

- **Device Provisioning Gateway (DPG)**: A device provisioning gateway is a software module. However, it can be provisioned in any local gateway or IoT hub. This is a minor implementation change only. This module performs three significant functions. Firstly, it receives the provisioning request from the device or the user on behalf of the device. Secondly, it interfaces with the policy database and fetches the appropriate pre-conditions that need to be satisfied before the device can be provisioned as well as the post-condition policies. Finally, it passes the respective policy information along with the device detail to the provisioning evaluation engine (PEE).
- **Provision Evaluation Engine (PEE)**: A PEE module is the decision making authority in this security architecture. The PEE module accepts all the device information and respective policy information from the DPG. After that, it evaluates the pre-conditions using the device attributes, parameters, and properties and determines whether the pre-condition policies are satisfied for any particular device. For any device, when the pre-conditions are completely satisfied, the PEE forwards the information to the DPS.
- **Device Provisioning Service (DPS)**: The DPS's purpose is to add the PEE-satisfied devices to the Provisioned Devices Database (PDD). This module also monitors the provisioned device activity and compares them with the post-condition policies. If any post-condition activity is not legitimate, then the DPS service can stop that particular device action. An event logger in conjunction with the DPS service logs the provisioned device activity. These event logs later help the security architecture to formulate a device's twin. In Section 4.5.4, we will describe the digital twin in detail.
- **PD**: The Policy Database stores the pre-condition and post-condition policies associated with the devices. This database can be implemented using JSON and can be mounted in the Azure IoT/AWS IoT. A local edge gateway or a hub can also be used to mount this database.
- **Post-P:** This database stores the device-specific post-condition policies. Each device can have multiple post-condition polices. The database is a simple JSON database.

4.5.3 SECURITY PROVISIONING PROTOCOL

Now we will explain the functional steps of the secure smart device provisioning module (Figure 4.8).

- **Step 1**: The device sends a provisioning request to the DPG.
- **Step 2**: The DPG retrieves two types of information: (i) device-specific and (ii) pre-/post-condition policies. With this information, DPG creates a device provisioning package (X). We have used Algorithm 2 to retrieve the information and create device provisioning package (x).

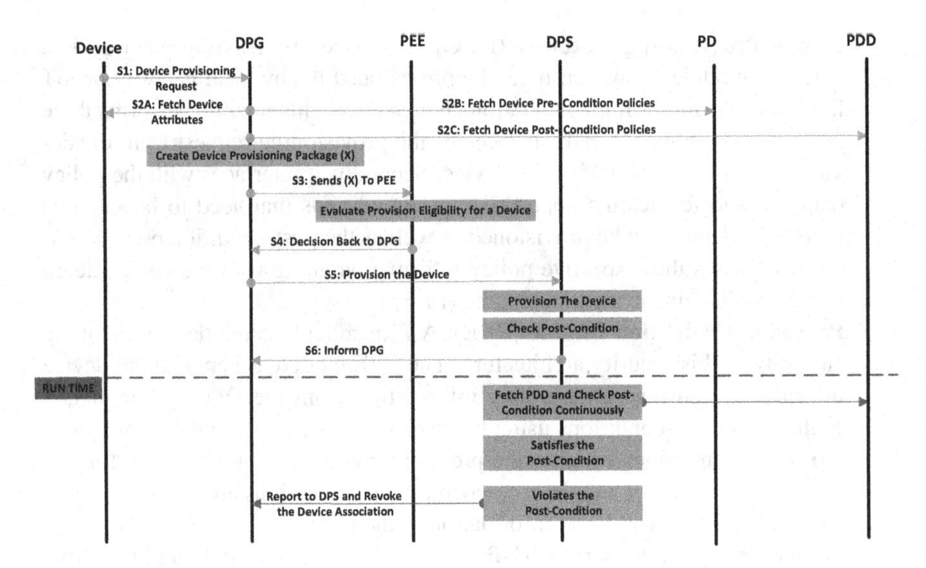

Figure 4.8 This figure describes how SDPM operates at a functional level. Here, we have presented the inter-module functional communication of the architecture components in both boot and runtime. Also, the figure demonstrates how pre- and post-conditions can interact with the sub-modules.

- **Step 3**: The DPG forwards the device provisioning package (X) to the PEE. The PEE compares the device attributes against the pre-condition policies, and it prepares a recommendation.
- **Step 4**: The PEE sends the recommendation back to the DPG.
- **Step 5**: If the device has satisfied the pre-conditions, then DPG requests the DPS to provision the device. If the device fails the pre-conditions, the DPG informs the device that it cannot be provisioned. The DPS registers the device, acts upon the post-conditions, and stores the results from the actions and any remaining post-conditions. That need to be executed during runtime in the POST-P database.
- **Step 6**: The DPS informs the DPG of the nature of the provisioning (successful/unsuccessful) of the device, which then informs the device accordingly.
- **Runtime**: During runtime, the DPS continuously checks the post-condition policies against the device's actions. If the actions are legitimate according to the post-condition policies, the DPS allows them to perform the action. Otherwise, the DPS stops them. Both types of events are logged in the event logger, which later helps the digital twin.

4.5.4 DIGITAL TWIN

In this section, we will explain the digital twin, its function, and its importance in SDPM.

Algorithm 2 Pseudo Algorithm for Device State Measurement

$TPM\_INFO, PS\_LIST, MEM\_REPORT, OS\_INFO$
$Provisioning\_Request$
$TI[] = TPM\_INFO$
$PL[] = PS\_LIST$
$MR[] = MEM\_REPORT$
$OI[] = OS\_INFO$
$PR = Provisioning\_Request$
while 1 **do**
  if $PR ==$ True **then**
    $fetch(TI[])$
    $TIH = hash(TI[])$
    $fetch(PL[])$
    $PLH = hash(PL[])$
    $fetch(MR[])$
    $MRH = hash(MRH[])$
    $fetch(OI[])$
    $OIH = hash(OI[])$
    $X = concat(TIH, PLH, MRH, OIH)$
    $send(X)$
  else
    $send(Device\ not\ compatible)$
  end if
end while

4.5.4.1 What Is a Digital Twin?

A digital twin is a digital representation of any provisioned device that is a part of the smart network infrastructure. A digital twin contains device data and meta-information, which defines a device. For instance, a temperature sensor logs a sensor reading for every 10 seconds. A digital twin of this particular sensor will also store the same data. Apart from that, a digital twin can store metadata and event-driven action associated with the data or the metadata. For instance, for our previous sensor scenario, it is possible to set an action to trigger an alarm when the temperature reaches a threshold value.

4.5.4.2 Digital Twin in SDPM

SDPM has a digital twin, which performs the following functions:

- It represents the smart devices in the smart network infrastructure.
- It stores the device event logs.
- It has granular policies over the event attributes. This facilitates the use of post-condition policies in the IoT network infrastructure.

4.5.4.3 How It Works?

In SDPM, the digital twin maintains a log of events or actions executed by the device at any point in time. For our digital twin, each event consists of action. Each action operates on a device state. Each device state is represented by device data, meta-data, and associated operation.

Each state of the device is represented as S_x, and each event is represented as e_y. A device state change from S_t to $S_{t'}$ for an event e_t at time t will be represented as:

$$S_t \xrightarrow{e_t} S_{t'} \tag{4.1}$$

So, the digital twin model for SDPM will be represented as:

$$\text{At time, } t :< S_t, e_t > \tag{4.2}$$

where, S_t = set of device state;
and e_t = set of events performed by the external and internal entities.

Here, we introduce event-driven policies. For instance, an industrial actuator uses a stepper motor that rotates at 200 rpm. The consecutive start duration is 10 minutes and the operational period is 2 minutes. However, an adversary has tampered with the device and changed the motor's speed, but not the operational duration. Our event-driven policies can detect this state change, and digital twin can enforce an action to raise an alarm or to stop the motor function to avoid damage.

Hence, our SPDM extends the digital twin concept by incorporating the device state into the meta-information and then monitoring this information to enforce specific policies for dynamically controlling device behaviour. This in turn is used to detect device vulnerabilities and achieve dynamic patch management in a secure manner. Furthermore, the pre- and post-provisioning policies used in our approach help to capture the pre-requisite authorisation requirements and post-obligation requirements in devices' behaviours.

4.5.5 POLICY-BASED SECURITY APPLICATION

Our programmable IoT network infrastructure consists of programmable network switches such as OpenFlow switches or IoT Gateways and end hosts (IoT sensors/actuators). A single-network controller manages the IoT network infrastructure. The programmable network devices (OpenFlow devices) forward the packets generated by the IoT devices/users, which are then subjected to the policies specified in the network controller for routing across the network. Figure 4.4 shows the policy-based security application (PbSA) for securing the programmable IoT network infrastructure. As *PbSA* is designed to be modular, the components of *PbSA* can be implemented on a single host or can be distributed over multiple hosts. Here, we provide a detailed description of different modules of *PbSA*. Modules are software components of the main application.

PbSA consists of four major modules, namely (i) Policy Manager, (ii) Evaluation Engine, (iii) Repositories, and (iv) Policy Enforcer.

- **Policy manager**: This is the core component of the security service application, as it manages every operation such as extracts IoT device flow attributes, updates the topology repository, and instructs the policy enforcer to enforce the policies at the OpenFlow IoT Gateways. It also communicates with the *IoT Provisioning* application for the transfer of authentication service tokens after checking the network service request from the IoT devices.
- **Evaluation engine**: The engine evaluates the service request against the relevant policies stored in the policy repository.
- **Repositories**: Our security service application has two repositories: (i) Topology repository and (ii) Policy repository. The topology repository contains the network topology of IoT devices and end hosts/users. Network controller might have its own device topology repository. We are using the same topology repository for this purpose. The policy repository contains the policy expressions (PEs) associated with the various IoT devices and the associated flow attributes. The attributes in PEs also include security parameters such as security labels associated with the programmable network IoT Gateways.

  Policy enforcer: The policy enforcer fetches the required information from the programmable IoT Gateways and enforces the routing rules or flow rules obtained from the policy manager.

4.5.5.1 Security Policy Specifications

The security policy specifications are expressed as policy expressions (PE), which specify whether packets and flows from IoT devices and end hosts follow a particular path or paths in the network, and the conditions under which the packets and flows follow these paths. The PE specification syntax uses an enhanced version of RFC1102 [11]. They are fine-grained and specify a range of policies using various attributes of IoT devices and flows; for instance, these attributes include different types of devices, source and destination attributes, flow attributes and constraints, requested services, security services, and security labels. The following attributes have been specified:

a. **Flow attributes**: Flow identifier, type of flow packets, security profile indicating the set of security services that are to be associated with the packets in the flow;
b. **Device attributes**: Identifiers specific to IoT sensor/actuator;
c. **Switch attributes**: Identifier of the switch and security labels associated with the switch (as well as OpenFlow IoT Gateways);
d. **Host attributes**: Identifiers associated with the host such as source/destination host ID;

e. **Fow and domain constraints**: Constraints such as flow constraints (Flow-Cons) and domain constraints (DomCons) associated with flows from a specific device; for instance, a constraint might specify the flow from a specific type of sensor, should only go through a set of switches that can provide a guaranteed bandwidth; from a security point of view, a constraint could be that a flow should only go through OpenFlow switches that have a particular security label;
f. **Services**: The services to which the PE applies (e.g., FTP storage access);
g. **Time validity**: The period for which the PE remains valid; and
h. **Path**: Indicates a specific sequence of switches that particular flows from specific IoT devices/users are allowed or should traverse. The PEs support wildcards for attributes, enabling the language to specify policies for a group of IoTs/services.

A simplified policy expression template is as follows:

$$PE_i = < FlowID, IoTDeviceID, SourceAS, DestAS, SourceHostIP, DestHostIP, SourceMAC, DestMAC, User, FlowCons, DomCons, Services, Sec-Profile, Seq-Path >:< Actions >$$

where i is the policy expression number.

4.6 PROTOTYPE IMPLEMENTATION

In this section, we will explain how we have created and implemented a proof-of-concept (PoC) prototype of the proposed security architecture.

4.6.1 NETWORK SETUP

We have developed a proof of concept for the proposed security architecture. The proof-of-concept implementation has network, hardware, and software modules. Some parts of it is implemented in the Azure cloud. Our implementation setup uses Oracle VM Box, Mininet-WiFi [14] and ONOS as the open-source network controller. We are using a workstation with Core i7 - 7700K @ 4.20 GHz CPU; 64 GB of RAM for this setup. Here, the local IoT hubs communicate with the Microsoft Azure IoT cloud services. The hub is running as a virtual machine, and the virtual IoT devices are connected to the hub. The Azure services collect the data sent by these devices. Our SDPM and PbSA services enforce the policies at the network infrastructure level and integrate with the core application services running on the Azure cloud. The proof-of-concept network configuration is shown in Figure 4.9 runs an ONOS controller inside a Ubuntu Server VM. The Mininet-WiFi is used to simulate a wireless network and Raspbian VM acts as IoT devices are connected to the WiFi network.

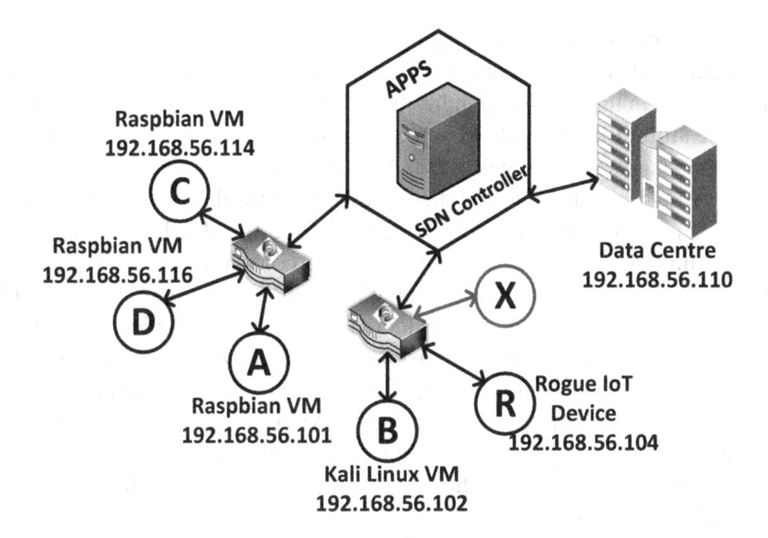

Figure 4.9 Network configuration.

4.6.2 SECURITY ANALYSIS

In this section, we will present first the security properties, then present some attack case studies. Finally, we will mould the attack case studies to real-time attack scenarios in which we will explain how our proposed SDPM provides security services to the devices and the network infrastructure.

4.6.2.1 Security Properties

Here, we present various security properties of SDPM. We have provided an empirical security feature evaluation of the SDPM. Here, we have explained the security capabilities of the SDPM and presented case studies to demonstrate these capabilities using real-time attacks. Later, we have discussed real attack scenarios that can be counteracted using the proposed SDPM.

- **Authentication**: SDPM offers fine-grained authentication services. The Microsoft Azure services offer X.509 and symmetric key-based authentication for IoT devices and hubs. We have extended the capabilities by introducing pre-conditioning policy-based authentication for IoT devices and hubs. SDPM pre-condition policies include user/device attributes and devices states. For instance, a device with untrusted processes running in it will never be provisioned. Hence, SDPM precondition policies help to authenticate the devices and provision them securely.
- **Authorisation**: As the IoT devices are resource constrained and may fail to meet optimum hardware requirements to run the security services, with SDPM, we have introduced post-condition policies. These policies help to monitor the state of the provisioned devices and their behaviour. The post-condition

policies consider device state and device/infrastructure specific contexts. For instance, with a post-condition, we can limit the sensor device packet transfer rates. This, in turn, can help in limiting DDoS attacks from malicious devices.

- **Device state**: Like any other computer system, IoT devices change their software and hardware state with time. Any attack towards these IoT devices may push them to an untrustworthy state, leaving both the device and networking infrastructure vulnerable to further attacks. Hence, state (hard/soft) of the devices is an important factor and should always be considered. Both pre- and post-condition policies consider the state of the devices. Any state change which is not reflected in the policies is considered as a suspicious device. In such cases, the device could either be unprovisioned or isolated for further investigation.
- **Digital twin**: In normal practice, a digital twin does not reflect the security properties of a device. However, with SDPM, we modelled the digital twin to reflect the security-specific aspects of the devices, for instance, devices states, environment contexts, etc. They help in enforcing post condition policies in the IoT network infrastructure.
- **Attack detection and mitigation**: SDPM is geared to detect attacks with pre- and post-condition policies. Also, with granular policy enforcement architecture, SDPM can un-provision, isolate devices, and mitigate attacks.
- **Flow authorisation**: Another important requirement is the need to be able to control and secure the flows between the IoT devices in the network. A major advantage of using programmable networks like SDN is its ability to provide domain-wide policy management for secure control of dynamic flows between IoT devices. Our policy-based security application (PbSA) enables fine-grained flow and path-based secure routing policies enforcing secure communication between end to end services and devices. For instance, a particular path can be restricted to only devices with a security label of at least high and a specified level of throughput while also being constrained to a set of specified (secure) paths. Such path-based policies are critical when securing data from sensitive devices but are also useful for applications with different quality of service requirements. For instance, traffic requiring certain bandwidth needs to take a path where the network devices and channels have the necessary capabilities. Furthermore, suppose due to some attack (e.g., DDoS attack), traffic from a device is not able to get through the network. Our PbSA provides an alternative path for the traffic from the device to reach its required destination. This highlights the novel feature of our PbSA to dynamically manage flow and path-based security policies to achieve secure communications across domains.
- **Flow Confidentiality, Integrity, and Availability (CIA)**: Our proposed architecture provides on-demand CIA services. These requirements are specified as part of the policy specifications in the PbSA module. In terms of confidentiality, specific flows along specific paths between specific devices can be encrypted. The encryption keys are established via a secure key management

process. The flows remain encrypted from end to end, and the intermediary OpenFlow switches will not be allowed to decrypt the payload. Similarly, the flows are also protected for integrity using cryptographic mechanisms. Moreover our security architecture ensures the devices are authenticated and their access to services and flows are determined by the access privileges granted to these devices as per the policy specifications in the PbSA.

4.6.2.2 Attack Case Studies

- **Case 1 - A device sending a single malicious packet**: In this scenario, an adversary steals the device credentials and launches an attack towards the IoT gateway.
  To defend against this attack, we need an authentication mechanism. We will simulate the attack with raspberry PI as a compromised IoT device.
- **Case 2 - A device sending a number of malicious packets**: In this attack scenario, a device floods the IoT gateway with a stream of fake packets. In this attack we want to test the robustness of the authentication services.
- **Case 3 - A compromised device injecting malware**: This scenario is very specific to devices that are authenticated and are already provisioned in the smart network infrastructure, for instance, devices infected with Mirai Malware. We will create attack scenarios of this type, and after that, we will test our authorisation policies.
- **Case 4 - A state change of a device due to external manipulation**: The digital twin provides state and event-driven monitoring features for a device. In this scenario, we will try to verify digital twin capabilities. Here, we will perform some attacks on the IoT devices that will cause state changes in the devices. These state changes are hard to detect by the authorisation policies. Our event-driven state monitoring services in digital twin can detect these changes. We will test such cases.
- **Case 5 - Spoofing/masquerading**: In the IoT network infrastructure, a malicious adversary can try to impersonate another user/device. An adversary can use such malicious approach to capture/modify the sensitive information from/in the devices as well as send malicious instructions to the actuators.
- **Case 6 - Man-in-The-Middle (MiTM) attack**: In MiTM attack, an adversary intercepts the communication between the two parties. The IoT infrastructure is also vulnerable to this type of attack. An adversary or a malicious IoT device can change the IoT device/user ARP caches of the two communicating parties to initiate a MiTM attack in the IoT network infrastructure.

Now, we will convert these attack case studies into real-time attack scenarios. Firstly, we will show, how an adversary can easily compromise an about-to-be provisioned device or already provisioned device. After that, we will show how SDPM can defend such attacks using pre-/post-condition policies.

4.6.2.2.1 *Attack Scenario 1*

This scenario is more specific to the attacks explained in case studies 1 and 2. As previously described earlier in these attacks, an adversary launches flooding towards the IoT device/hub/gateway/server.

In Figure 4.10a, a typical IoT environment is shown where multiple sensors collect real-time data and upload it to the cloud via a central IoT Hub. The cloud is used for data storage and analysis and reporting by other applications and services. In this attack scenario, an adversary gains access to one of the IoT devices and runs malicious code, which is then used to target and disable the upstream IoT hub. Here, X.509 certificates are used to the authenticate the devices with the IoT hub. We assume that the adversary has already captured the X.509 certificates. in early 2021, in SolarWind attack the adversaries captured the X.509 certificates of applications by modifying the Azure Active Directory settings [16]. Later, they used the keys to maintain access to the compromised network infrastructure and remained dormant for further attacks.

The attack takes the form of a network denial of service; in this case, we are using LOIC (Low Orbit Ion Cannon) to launch network packet flood [5] and overwhelm the upstream devices on the network. LOIC interface and settings are shown in Figure 4.10b, and we are running LOIC from a Kali Linux VM. As the attacker

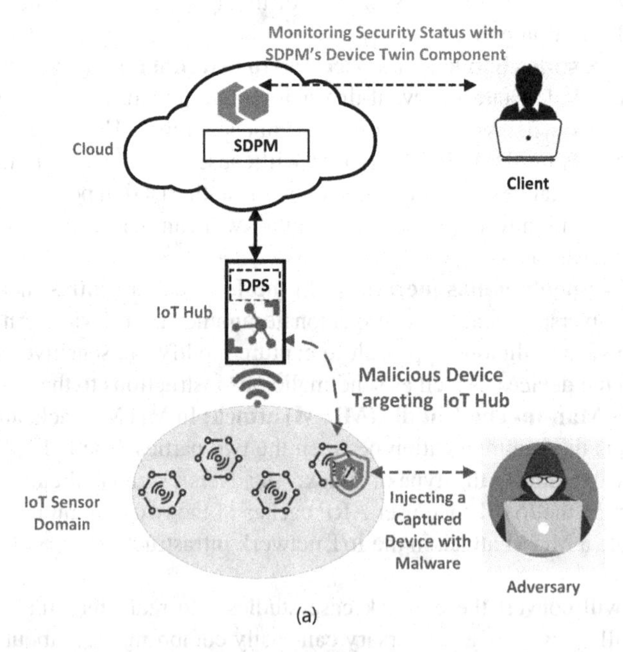

Figure 4.10 This figure presents an attack scenario (1). (a) A schematic diagram of the network setup is presented here. Also, the schematic shows how an adversary can penetrate an IoT sensor domain.

(Continued)

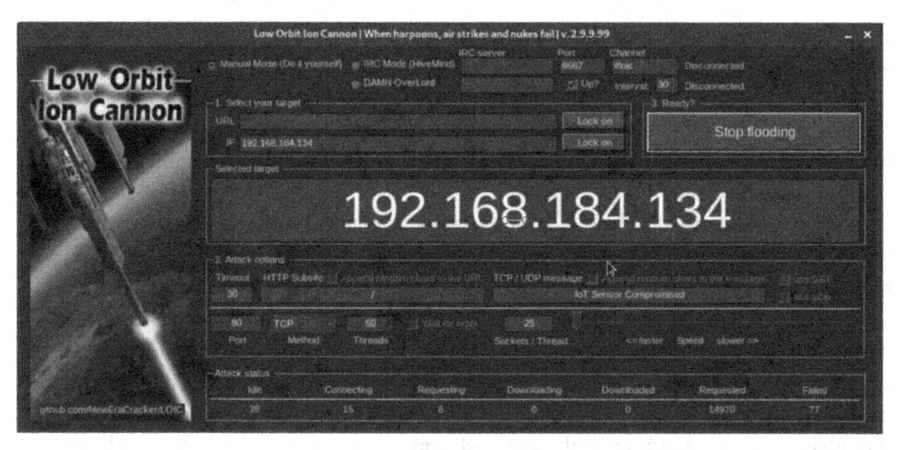

(b)

Apply a display filter … <Ctrl-/>

No.	Time	Source	Destination	Protocol	Length	Info
5683…	878.160264478	192.168.184.131	192.168.184.134	TCP	88	53854 → 80 [PSH, ACK] Seq=3521 Ack=1 Win=64256 Len=22 TSval=382274508 TSecr=15142911
5683…	878.160277396	192.168.184.131	192.168.184.134	TCP	88	53848 → 80 [PSH, ACK] Seq=3675 Ack=1 Win=64256 Len=22 TSval=382274508 TSecr=15142911
5683…	878.160300177	192.168.184.131	192.168.184.134	TCP	88	53938 → 80 [PSH, ACK] Seq=23 Ack=1 Win=64256 Len=22 TSval=382274508 TSecr=151429114
5683…	878.160309761	192.168.184.131	192.168.184.134	TCP	88	53880 → 80 [PSH, ACK] Seq=2355 Ack=1 Win=64256 Len=22 TSval=382274508 TSecr=15142911
5683…	878.160395856	192.168.184.131	192.168.184.134	TCP	88	53878 → 80 [PSH, ACK] Seq=2267 Ack=1 Win=64256 Len=22 TSval=382274508 TSecr=15142911
5683…	878.160480302	192.168.184.131	192.168.184.134	TCP	88	53908 → 80 [PSH, ACK] Seq=1321 Ack=1 Win=64256 Len=22 TSval=382274508 TSecr=15142911
5683…	878.160517723	192.168.184.134	192.168.184.131	TCP	66	80 → 53910 [ACK] Seq=1 Ack=1255 Win=28992 Len=0 TSval=1514291149 TSecr=382274506
5683…	878.160517766	192.168.184.134	192.168.184.131	TCP	66	80 → 53890 [ACK] Seq=1 Ack=2179 Win=28992 Len=0 TSval=1514291149 TSecr=382274506
5683…	878.160517796	192.168.184.134	192.168.184.131	TCP	66	80 → 53844 [ACK] Seq=1 Ack=4005 Win=28992 Len=0 TSval=1514291149 TSecr=382274506
5683…	878.160517822	192.168.184.134	192.168.184.131	TCP	66	80 → 53872 [ACK] Seq=1 Ack=2575 Win=28992 Len=0 TSval=1514291149 TSecr=382274506
5683…	878.160517854	192.168.184.134	192.168.184.131	TCP	66	80 → 53934 [ACK] Seq=1 Ack=353 Win=28992 Len=0 TSval=1514291149 TSecr=382274507
5683…	878.160517884	192.168.184.134	192.168.184.131	TCP	66	80 → 53854 [ACK] Seq=1 Ack=3521 Win=28992 Len=0 TSval=1514291149 TSecr=382274506
5683…	878.160517915	192.168.184.134	192.168.184.131	TCP	66	80 → 53900 [ACK] Seq=1 Ack=1805 Win=28992 Len=0 TSval=1514291149 TSecr=382274506
5683…	878.160517945	192.168.184.134	192.168.184.131	TCP	66	80 → 53908 [ACK] Seq=1 Ack=1321 Win=28992 Len=0 TSval=1514291149 TSecr=382274506
5683…	878.160540936	192.168.184.134	192.168.184.131	TCP	66	80 → 53856 [ACK] Seq=1 Ack=3411 Win=28992 Len=0 TSval=1514291149 TSecr=382274506
5683…	878.160540975	192.168.184.134	192.168.184.131	TCP	66	80 → 53870 [ACK] Seq=1 Ack=2443 Win=28992 Len=0 TSval=1514291149 TSecr=382274507
5683…	878.160541007	192.168.184.134	192.168.184.131	TCP	66	80 → 53928 [ACK] Seq=1 Ack=463 Win=28992 Len=0 TSval=1514291149 TSecr=382274507
5683…	878.160541045	192.168.184.134	192.168.184.131	TCP	66	80 → 53862 [ACK] Seq=1 Ack=3081 Win=28992 Len=0 TSval=1514291149 TSecr=382274507
5683…	878.160541078	192.168.184.134	192.168.184.131	TCP	66	80 → 53930 [ACK] Seq=1 Ack=419 Win=28992 Len=0 TSval=1514291149 TSecr=382274507
5683…	878.160541111	192.168.184.134	192.168.184.131	TCP	66	80 → 53882 [ACK] Seq=1 Ack=2289 Win=28992 Len=0 TSval=1514291149 TSecr=382274507
5683…	878.160541144	192.168.184.134	192.168.184.131	TCP	66	80 → 53940 [ACK] Seq=1 Ack=23 Win=28992 Len=0 TSval=1514291149 TSecr=382274507
5683…	878.160541173	192.168.184.134	192.168.184.131	TCP	66	80 → 53876 [ACK] Seq=1 Ack=2407 Win=28992 Len=0 TSval=1514291149 TSecr=382274507
5683…	878.160567204	192.168.184.134	192.168.184.131	TCP	66	80 → 53866 [ACK] Seq=1 Ack=2641 Win=28992 Len=0 TSval=1514291149 TSecr=382274507
5683…	878.160567239	192.168.184.134	192.168.184.131	TCP	66	80 → 53886 [ACK] Seq=1 Ack=2311 Win=28992 Len=0 TSval=1514291149 TSecr=382274507
5683…	878.160567262	192.168.184.134	192.168.184.131	TCP	66	80 → 53896 [ACK] Seq=1 Ack=2025 Win=28992 Len=0 TSval=1514291149 TSecr=382274507
5683…	878.160567291	192.168.184.134	192.168.184.131	TCP	66	80 → 53858 [ACK] Seq=1 Ack=3213 Win=28992 Len=0 TSval=1514291149 TSecr=382274507
5683…	878.160567317	192.168.184.134	192.168.184.131	TCP	66	80 → 53912 [ACK] Seq=1 Ack=1189 Win=28992 Len=0 TSval=1514291149 TSecr=382274507
5683…	878.160567344	192.168.184.134	192.168.184.131	TCP	66	80 → 53918 [ACK] Seq=1 Ack=881 Win=28992 Len=0 TSval=1514291149 TSecr=382274507
5683…	878.160567366	192.168.184.134	192.168.184.131	TCP	66	80 → 53914 [ACK] Seq=1 Ack=1057 Win=28992 Len=0 TSval=1514291149 TSecr=382274507
5683…	878.160567398	192.168.184.134	192.168.184.131	TCP	66	80 → 53926 [ACK] Seq=1 Ack=485 Win=28992 Len=0 TSval=1514291149 TSecr=382274507
5683…	878.160615795	192.168.184.134	192.168.184.131	TCP	66	80 → 53902 [ACK] Seq=1 Ack=1607 Win=28992 Len=0 TSval=1514291149 TSecr=382274507
5683…	878.160615838	192.168.184.134	192.168.184.131	TCP	66	80 → 53922 [ACK] Seq=1 Ack=859 Win=28992 Len=0 TSval=1514291149 TSecr=382274507
5683…	878.160615865	192.168.184.134	192.168.184.131	TCP	66	80 → 53916 [ACK] Seq=1 Ack=1035 Win=28992 Len=0 TSval=1514291149 TSecr=382274507
5683…	878.160615892	192.168.184.134	192.168.184.131	TCP	66	80 → 53864 [ACK] Seq=1 Ack=2839 Win=28992 Len=0 TSval=1514291149 TSecr=382274507

(c)

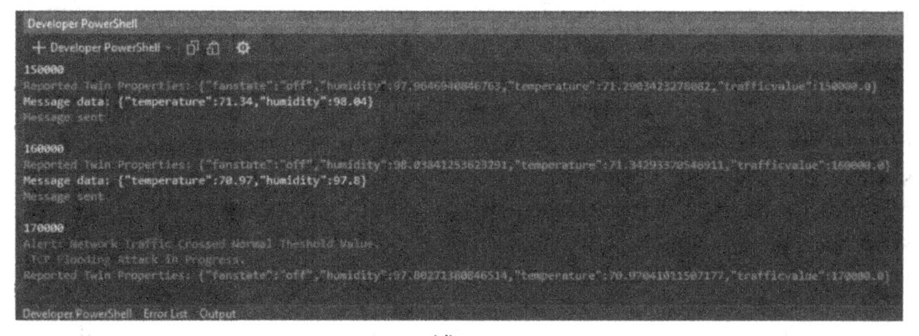

(d)

Figure 4.10 ***(Continued)*** (b) Low orbit ion canon [5] application is used to generate the flooding attack. The tool configuration to launch the attack is presented in the figure. This figure presents the after attack artefacts from scenario 1. (c) The figure presents the wireshark trace while the attack is happening. (d) This figure presents how SDPM generates an alert for an ongoing flooding attack.

compromised device floods the hub, it tries to process the packet request, which consumes a huge amount of memory and processing power. Finally, the attack proceeds to the point where the level of incoming traffic reaches a peak and then the hub freezes. We have shown a wireshark trace of the attack in Figure 4.10c. We have also noticed that, it takes about 4 minutes from the time the attack is launched for the peak to occur and the hub freezes.

Now we will activate SDPM, which uses post-conditions and policies to mitigate such Denial of Service attacks. These post-conditions work after a device is provisioned and connected. Basically, these post-conditions monitor the device behaviour after it is provisioned and check if the device is behaving maliciously. If so, this is stopped. With SDPM we achieve this via the notion of a digital twin. As described previously, a digital twin is a digital representation of the physical IoT device and replicates the physical device's state attributes. So in this case, the digital twin has the attribute corresponding to the network traffic value sent by the physical device. Now the monitoring service can check the digital twin's state and check whether the network traffic value exceeds the safe threshold, previously set by the policies. This is what the post-conditions do. If the safe threshold is exceeded, then the physical device is disconnected. For the current attack, our SDPM monitoring service continually queries for any device twins whose traffic value exceeds 150,000. Figure 4.10d shows that, when the traffic value grows beyond 150,000, the monitoring service sends an alert for the flooding attack and SDPM blocks the device. Thus, SDPM secures the IoT infrastructure from flooding attacks.

4.6.2.2.2 Attack Scenario 2

This attack is more specific to preconditioning checks before the device gets provisioned (specific to case study 3 and 4). These checks are done prior to provisioning, and we will show how these checks can help improve security with this scenario.

In this scenario, we consider that the sensor is purchased from a manufacturer or via an online store, and it is provisioned and connected to the cloud via a central IoT hub for communications, delivery of telemetry data, and management. Now we consider the situation where an adversary compromises an IoT sensor. For instance, it contains malicious code. This can happen if the seller is a bad actor. That is, a bad guy, an adversary, modifies the IoT device by installing malicious code before selling it on the internet to unsuspecting customers. In this scenario, we will consider a typical Raspberry Pi device, which was compromised previously by a malware script. Later, when the current owner tried to provision the device, it downloads and launches a ransomware attack. Figure 4.11b shows the active ransomware running in the raspberry device. The wireshark trace in Figure 4.11c shows that the malware is active and connects to a ransomware site – ransomeware.com. After connecting to the site, it tries to download the ransomware payload file – file.exe, which can then be used to launch the ransomware attack.

Now, to mitigate such propagation of malware and stopping and detecting such compromised devices, we will use the pre-conditioning policies in SDPM. To be specific, we will be using the device state properties in the pre-conditioning policies. This is an extension to the traditional way of provisioning the devices with just ID and certificates. Such granular device state-specific policies give us more flexibility to control the device, which helps mitigate attacks discussed in this scenario.

We have activated the SDPM services in our IoT network infrastructure. We have policies associated with the device state of the Raspberry Pi devices. In this case, we added the processes running in the Raspberry Pi devices as the state information in the pre-condition policy. At first, the device gets its attributes by loading its device

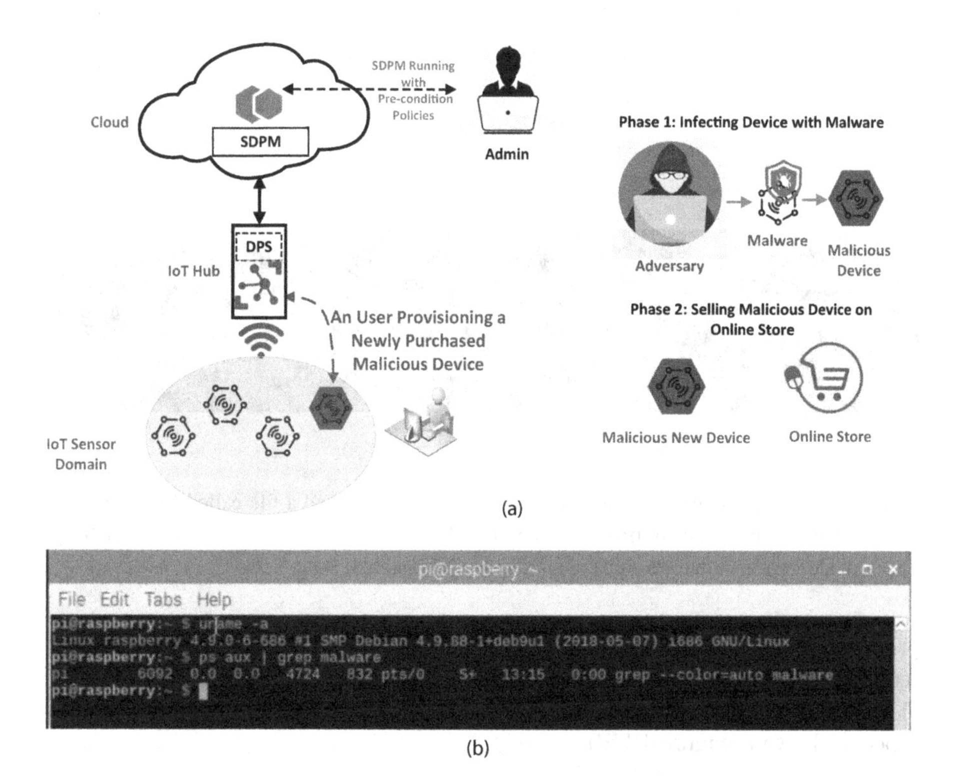

Figure 4.11 The figure describes the network setup for attack scenario 2. In this specific attack, a device is infected with malware, and SDPM is monitoring its status change due to malware infection. (a) The schematic description of how the attack is conducted. (b) This figure presents an active malware process inside a Raspberry Pi device. This malware is specifically designed for the IoT devices.

(Continued)

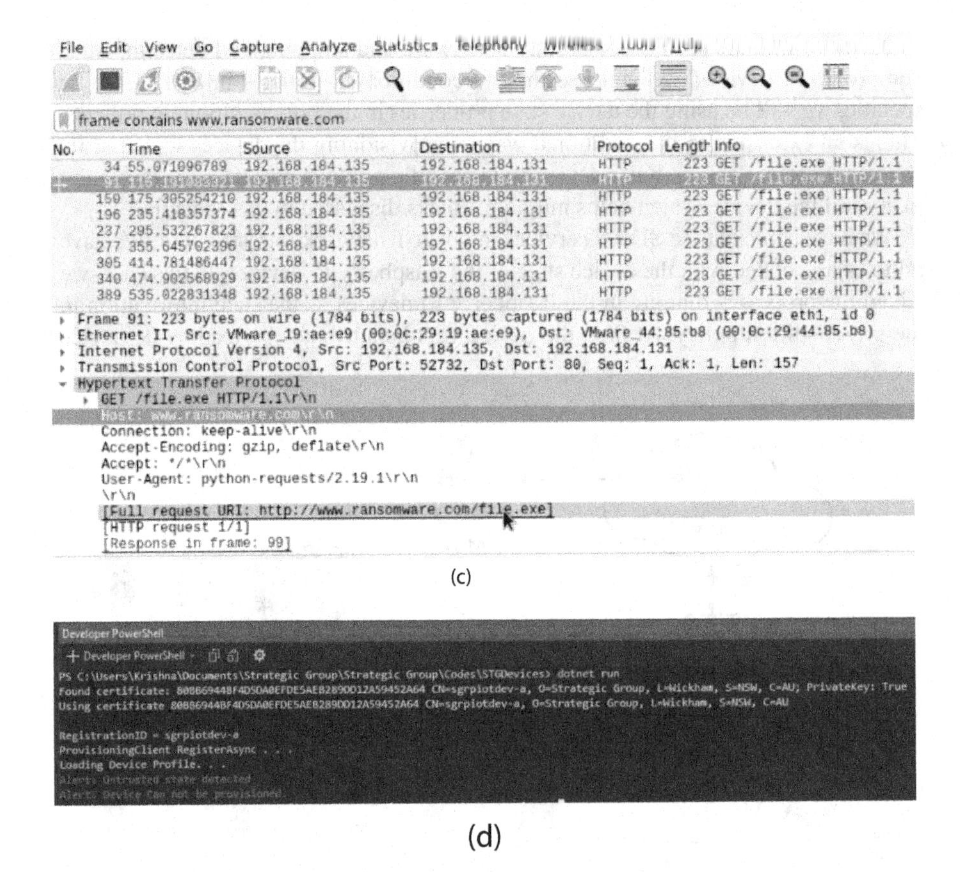

Figure 4.11 ***(Continued)*** (c) The malware will try to download a file called File.exe. The figure presents the Wireshark trace for the infected device requesting file.exe. (d) The SDPM raises an alert while it detects a change in the device's internal state.

profile and then checks the pre-conditions. As the malware process and the network request generated by the devices violate the state properties in the precondition policies, SDPM generates an ALERT. This leads to the cancellation of the provisioning process (shown in Figure 4.11d).

4.6.2.2.3 Attack Scenario 3

Case study 5 & 6 are interlinked with each other. Here, in order to launch a MiTM attack (Case 6), we have used ARP cache poisoning of host user/devices. The ARP cache poisoning uses ARP spoofing (Case 5).

These two attacks are typical of an insider attack in any IoT infrastructure, where an employee sensor data for malicious purposes [42]. Firstly, we will present how the attacks work, and then, we will show how our security architecture helps to protect the network.

We have used one of the Raspberry VMs as an IoT device and have used a web-server (IP: 192.168.56.101) to log the IoT data. The web-server has a dedicated user/-password for the admin user and devices. In this case, it is: "admin/password". The IP address of the Kali Linux VM is 192.168.56.102. The adversary uses SDNPWN Toolkit to poison the device/user ARP caches using *arpspoof* attack (Threat 3) [36,47]. To poison the ARP cache, the adversary first sends a legitimate flow request to the controller to install a flow rule, which establishes a route to the victim IoT device. Then, the adversary forges a gratuitous ARP request to poison the APR cache of the IoT device. The adversary also poisons the ARP cache of another user. The IoT device packets are rerouted through the malicious adversary. This essentially allows the adversary to view, copy, and modify any instruction/data to or from the IoT device. The adversary is also able to launch a MiTM attack (Case 6). Moreover, it introduces the possibility of additional attacks such as SSL stripping and session hijacking. Figure 4.12a shows a successful attack. The device ARP cache poisoning reroutes the device traffic through malicious adversary allowing him to capture the log-in information to the Raspbian web-server. Hence, the adversary is able to successfully launch an MiTM attack and capture the log-in details. Figure 4.12b shows the administrator log-in request containing the login ("admin") and password ("password") that have been captured.

With our proposed security architecture, each of the devices, as well as the user machines, is securely provisioned before it is connected to the IoT infrastructure. An adversary, who is not an insider, will not be able to bypass the authentication phase. On the contrary, if the adversary is an insider and is able to pass the provisioning phase, then the adversary will be restricted by the security policies in the network flow authorisation phase using the PbSA module. Also, at the end of the authentication phase, each IoT device will establish a secret key, which is used to encrypt its flows. This creates a secure channel which the IoT device can use to send the credentials to the web-server. Hence, the adversary is unable to steal the credentials. Figure 4.12c shows the attempted attack detected by our architecture.

4.6.3 PERFORMANCE EVALUATION

In this section, we will present some performance data for our proposed solution. We classify the performance into two categories: system/application specific and device specific.

Our proposed security architecture may cause delays in the provisioning and functioning of the IoT devices. Hence, we present the following data: (i) Device

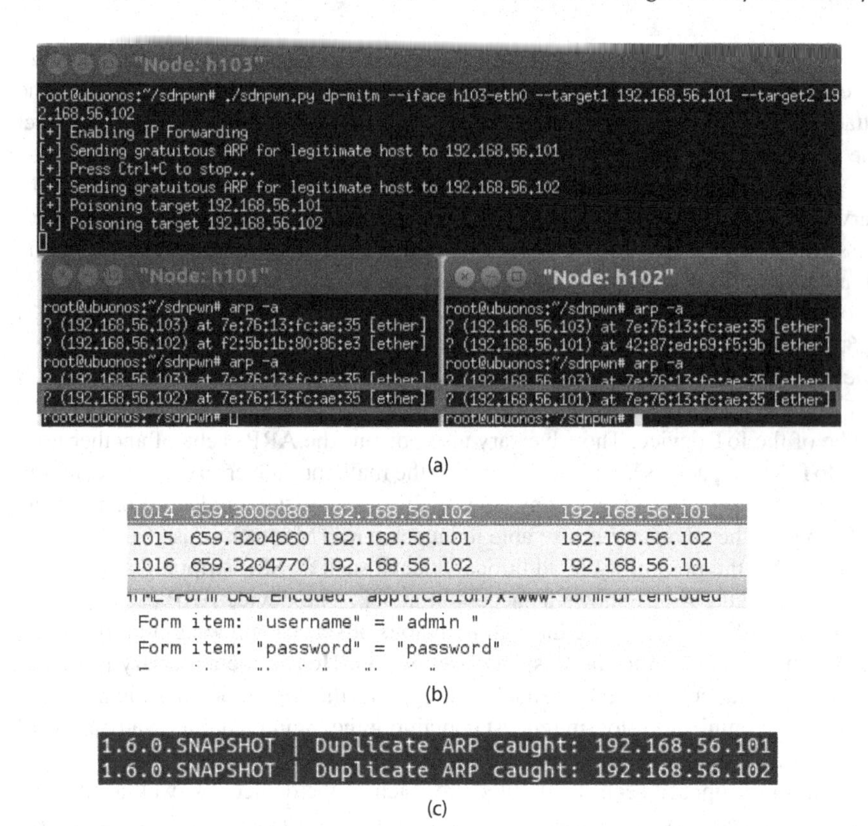

Figure 4.12 (a) This image 7 in a the attack scenario. Here, node:h103 is the attacker terminal and node:h101 and h102 are the victims of MiTM attack. Different IP allocated to the same MAC address in the the victim terminal presents a successful MiTM attack. (b) A packet capture from the adversary device presents the Admin/Password in bare form. (c) Our proposed tool successfully detects the MiTM attack.

provisioning time delay, (ii) Throughput, (iii) Resource (CPU/ Memory) utilisation, (iv) Device power consumption, and (v) Average path setup time.

We created a sample network with four hubs connected to the Azure cloud, and we have provisioned some dummy devices with each hub during each experiment. In this chapter, we focus only on the security issues of an IoT network infrastructure rather than other performance characteristics. Hence, increases in delay time and increased use of system resources are acceptable in the interests of increased security. Also, measurements presented in this section are very context specific and may vary in other environmental contexts.

i. **Device provisioning time delay**: Device provision time is the time required to provision an IoT device into an IoT network infrastructure automatically. This process consists of authentication, pairing, and authorisation processes.

As we have enhanced the capabilities of the authentication procedure by introducing pre-condition policies, there will be a slight delay in the overall device provisioning process. Figure 4.13a shows the provision time with changing the number of devices and the pre-condition policies associated with the authentication process. We measured the device provisioning time without SDPM, and we found that the provision time increases with the number of devices. For 50 devices, it takes 138 seconds, and it increases to 581.7 seconds with 200 devices. Now, after activating the SDPM security services, due to processing check, the provision time increases. With 200 devices and 300 pre-condition policies, the provision time is 1477.9 seconds.

ii. **Throughput**: Figure 4.13b shows a comparison between throughput with or without running the SDPM service for the IoT infrastructure. We varied the number of post-condition policies and measured the network infrastructure throughput using iperf. The bar chart illustrates a decrease in throughput with increasing post-condition policies. For example, with 300 policies, the throughput is 430 Mbps (appro.). While without the SDPM, the throughput is 840 Mbps(appro.).

iii. **Resource (CPU/memory) utilisation**: Figure 4.13c presents the system resources consumed by the SDPM services while being active. Here, we focus only on CPU usages and heap memory usages to describe the system resources usages. The resource utilisation is primarily dependent on the system used and environmental context during the execution time. Hence, this is not a standard representation rather a way of showing how a prototype system behaves. Here, we have conducted the experiments 10 times and used the average of the resources. The bar diagram shows that the CPU and heap memory usage increases with the higher number of policies (both pre- and post-conditions). With 400 policies, the CPU usages become 84%, and the software utilised 1,120 MB of heap memory. On the contrary, without the SDPM services running on the IoT network hubs, they utilise on an average 45% of the CPU and 129MB heap memory.

iv. **Device power consumption**: We present the device end-to-end average path setup time (with varying OF-AP) and battery consumption of an active IoT device. In Figure 4.13d with default applications running in the IoT infrastructure, the IoT device's battery decreases steadily to 73% within an hour of communication. With PbSA and ISA, the IoT device's battery consumption decreases steadily to 67% within an hour. The security mechanisms used in the IoT devices consumes their battery power, which is visible in 4.13.

v. **Average path setup time**: The average path setup time between IoT devices increases in both cases with the varying number of OF-AP (Figure 4.13e). With ONOS default applications running, the average path setup time for 100 OF-AP is 330.66 ms, and it steadily increases to 2487.68 ms with 500 OF-APs. On the contrary, PbSA and ISA running over the controller cause some delays in path setup time. In this case, with 100 OF-AP, the average path setup time is 349.76 ms, and it steadily increases to 2589.18 ms with 500 OF-AP.

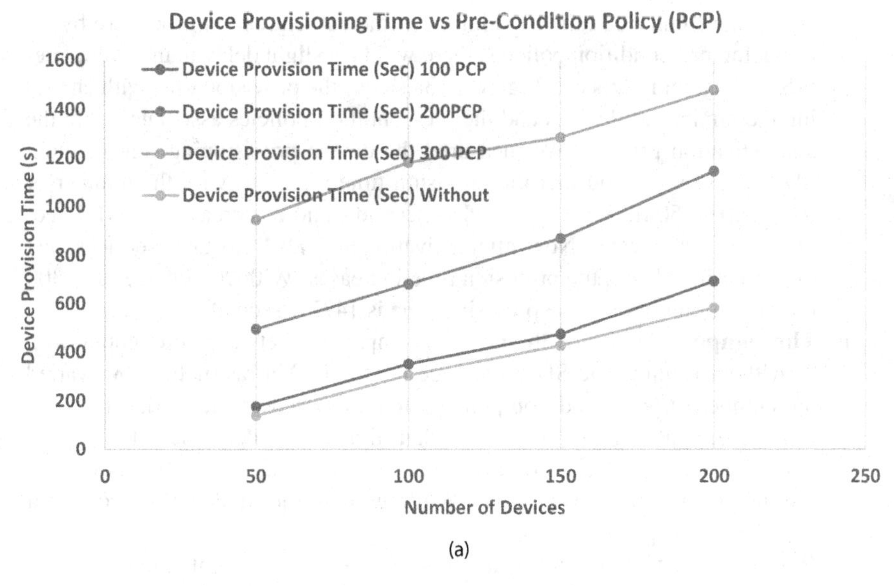

(a)

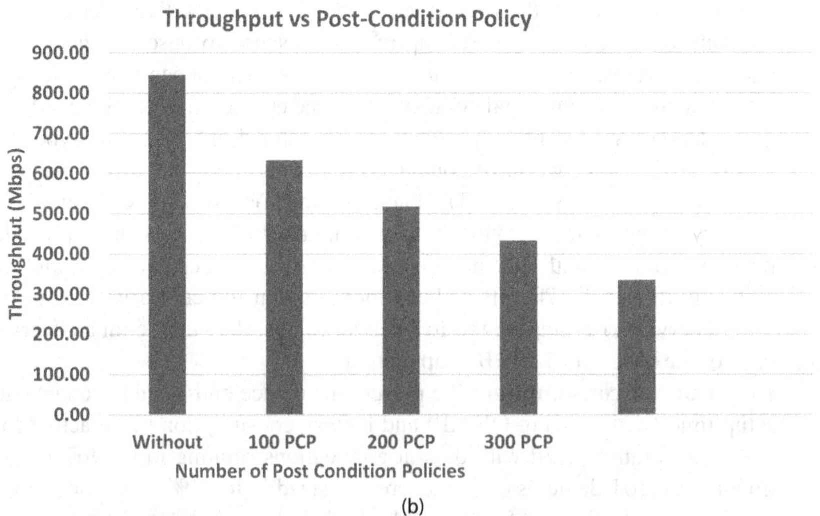

(b)

Figure 4.13 (a) The line chart presents the device provisioning time with varying number of Pre-Condition Policy (PCP). Here, we see an increase in the device provision time with increasing number of devices and PCP. (b) This bar chart presents the decrease in throughput with the increase in PCP per device.

(Continued)

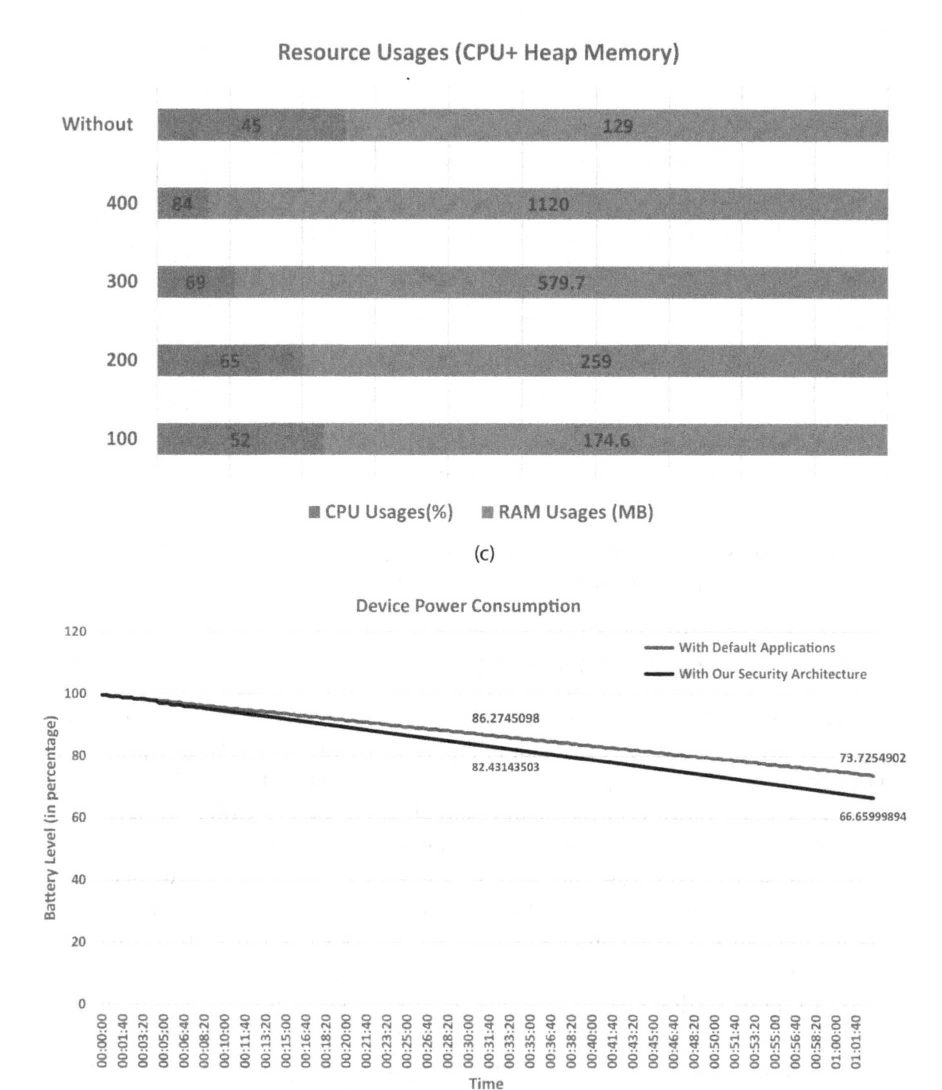

Figure 4.13 ***(Continued)*** (c) This graph shows consumption of CPU and heap memory resources while SDPM is active. (d) The figure presents a single simulated device's power consumption while using and/or interacting with the proposed security architecture. Our proposed solution discharges the IoT device faster than the normal one.

(Continued)

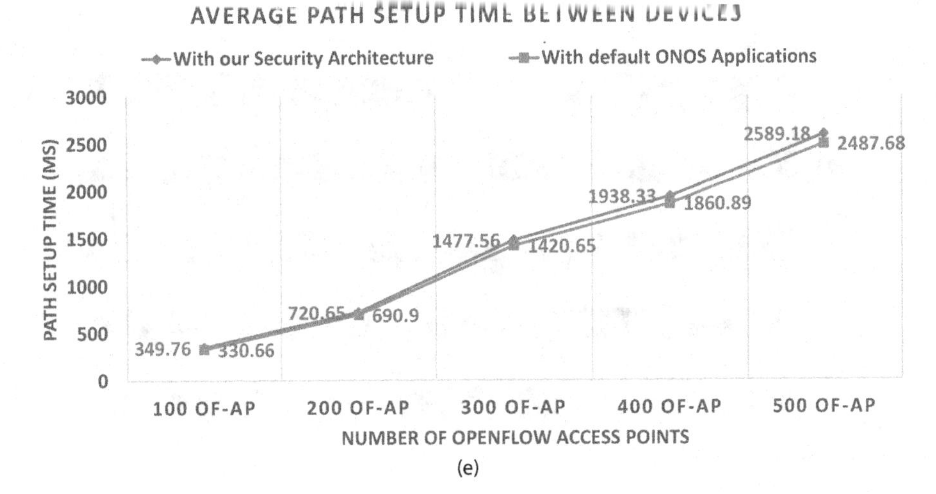

Figure 4.13 *(Continued)* (e) This figure presents the average path setup time between devices. We have varied the access point numbers and collected the setup time.

4.7 DISCUSSION AND OPEN ISSUES

IoT devices are increasing day by day, and they are manufactured for domains or a particular task. For instance, medical IoT devices try to help healthcare providers achieve the body vitals of a patient. Each domain-specific IoT device comes with its own protocols, network infrastructure, cloud application, security services, etc. A flaw in any of these services and infrastructure can lead to disruption and leakage of personal and corporate information. In order to safeguard the devices and network assets, we proposed this consolidated security architecture. Our primary focus for this work is to securely provision the IoT devices and securely manage programmable IoT network infrastructure. The proposed approach is different from the works present in this research domain. Firstly, in addition to traditional device authentication (while provisioning the device), we can introduce additional security checks using the granular device and context-specific pre-condition policies. Secondly, the SDPM service in the security architecture has pre-condition policies to cross-check the device operation during runtime (authorisation). The pre-condition policies are created around the digital twin concept. We have created security feature-focused digital twins, which allow us to check the security status of the provisioned devices. Thirdly, we consider device state as one attribute for pre- and post-condition policies. This allows the security architecture to check the device's security status both in the provision and runtime. For instance, whether the device is running a malware service or the device firmware is outdated. Fourthly, the security architecture can block malicious devices from getting provisioned in the network infrastructure and stop adversaries' attacks. Finally, the security architecture uses granular security policies to route IoT device flows securely. The security architecture provides on-demand security services such as confidentiality and integrity to users and devices. Table 4.1 shows a feature comparison between our work with [17,20,43,44].

Table 4.1
Security Feature Comparison of SDPM

Features	Additional Pre-condition Policy for Device Authentication	Digital Twin Associated Post-Condition Policy for Device Authorisation	Device State as a Policy Attribute	Network Security Management Policies	Attack Mitigation
Our Approach	Yes	Yes	Yes	Yes	Yes
[17]	No	No	No	No	No
[43]	No	No	No	No	No
[44]	No	No	Yes	No	No
[20]	No	No	No	No	No

Some open issues need some clear attention. Firstly, optimisation of security services for IoT devices and associated network infrastructure. This includes efficient and lightweight authentication and authorisation techniques for device provision. These techniques may vary depending of the type of IoT device and network configuration. Secondly, mass updates or patching of IoT devices is a significant concern and needs some focus. Thirdly, policy conflict is prevalent in the policy-based management system. Hence, it is an area that needs more exposure. Finally, future programmable networks will introduce new vulnerabilities into IoT services. So, we need to assess such avenues.

4.8 CONCLUSION

This chapter presents how policy-driven architectures can be used to manage and provision IoT devices and network infrastructures securely. It also introduces standard policy-based security architecture and its components. Such standards help us put security services and measures into the right environment. Here, we have added policy-driven services to the IoT network infrastructure. We have introduced two types of security services in this architecture: a policy-based secure provisioning service for onboarding IoT devices and a policy-based security application for secure IoT network infrastructure management.

The first security service is named SDPM, a policy-based approach for the secure provisioning of IoT devices. Here, we have presented SDPM service that can automate secure provisioning of IoT devices using cloud infrastructure. SDPM utilises pre- and post-condition policies related to sensor and actuator attributes and device state properties, which can be used to enforce specific security constraints while provisioning devices. They can also use device adapter-specific policies. The primary purpose of such device adapters is to extract and provision sensor data to IoT applications and to receive control commands from them. The proposed approach can be extended to include policies for device adapters specifying what sort of communications and what type of scripts are allowed for a device or a set of devices. For instance, this can include policies specifying access to either specific hosts and applications for cloud-based services or even certain classes for access within an operational network. An example of a class policy might be to "allow access to devices of the same manufacturer". We can also have policies on the type of communication protocols that are allowed for a device or set of devices at a certain location. For instance, "devices located in a specific domain should only communicate with IPSec". The SDPM service uses a digital twin representation of a device to reflect the security state of the devices during runtime. We create the digital twin of a device using the device information and devices/user/environmental context information. The SDPM has policies associated with such context attributes. This digital twin representation of the device can then be synchronised with the physical device. This will provide a starting point for secure monitoring of the health of the devices using cloud-based security applications. The digital representations help the users to get feedback about misbehaving or faulty devices in the environment. Furthermore, we developed a prototype application for SDPM using Microsoft Azure. This cloud-based security application enables secure access to sensor data of devices and secure control of the devices, allowing precise control over the smart infrastructure. Both pull-based and push-based approaches can be done. Here, we use a pull-based approach, where SDPM actively retrieves sensor data based on a send/request model.

A policy-based security application is another policy-driven security service that we have introduced as part of the security architecture. The PbSA helps to secure the IoT network infrastructure. This application provides two major benefits: firstly, to secure the network assets from potential compromises, and secondly, to provide on-demand security services to IoT devices and user flows. For example, a medical IoT device sending privacy critical information to the cloud demands confidentiality services.

We present a prototype implementation of the security architecture using Microsoft Azure and SDN controllers. We tested the proposed SDPM and PbSA services against some well-known attacks and demonstrated the outcome. The proposed approach applies to provisioning IoT devices and management of IoT network infrastructure in various applications such as agricultural farm, healthcare, and smart home environments.

REFERENCES

1. Mohammad Aazam and Eui-Nam Huh. Dynamic resource provisioning through fog micro datacenter. In *2015 IEEE International Conference on Pervasive Computing and Communication Workshops (PerCom Workshops)*, pp. 105–110, St. Louis, MO. IEEE, 2015.
2. Manos Antonakakis, Tim April, Michael Bailey, Matt Bernhard, Elie Bursztein, Jaime Cochran, Zakir Durumeric, J. Alex Halderman, Invernizzi, Michalis Kallitsis, et al. Understanding the mirai botnet. In *26th {USENIX} Security Symposium ({USENIX} Security 17)*, pp. 1093–1110,Vancouver, BC. 2017.
3. Lei Bai, Lina Yao, Salil S. Kanhere, Xianzhi Wang, and Zheng Yang. Automatic device classification from network traffic streams of internet of things. In *2018 IEEE 43rd Conference on Local Computer Networks (LCN)*, pp. 1–9, Chicago, IL. IEEE, 2018.
4. Soma Bandyopadhyay, Munmun Sengupta, Souvik Maiti, and Subhajit Dutta. A survey of middleware for internet of things. In Abdulkadir Özcan, Nabendu Chaki, Dhinaharan Nagamalai (Eds.), *Recent Trends in Wireless and Mobile Networks*, pp. 288–296. Springer, Berlin, 2011.
5. A. M. Batishchev. Loic (low orbit ion cannon). *online] http://sourceforge. net/projects/loic*, 2004.
6. Olaf Bergmann, Kai T. Hillmann, and Stefanie Gerdes. A coap-gateway for smart homes. In *2012 International Conference on Computing, Networking and Communications (ICNC)*, pp. 446–450, Maui, HI. IEEE, 2012.
7. Robert Bogue. Towards the trillion sensors market. *Sensor Review*, 34(2):137–142, 2014.
8. Xianghui Cao et al. Ghost-in-ZigBEE: Energy depletion attack on zigbee-based wireless networks. *IEEE Internet of Things Journal*, 3(5):816–829, 2016.
9. Marc Capellupo et al. Security and attack vector analysis of IoT devices. In *International Conference on Security, Privacy and Anonymity in Computation, Communication and Storage*, pp. 593–606, Guangzhou. Springer, 2017.

10. Abdur Rahim Choudhary. Policy-based network management. *Bell Labs Technical Journal*, 9(1):19–29, 2004.
11. D. D. Clark. Policy routing in internet protocols request for comment RFC-1102. Network Information Center, 1989.
12. Ana Cristina Franco da Silva, Pascal Hirmer, Jan Schneider, Seda Ulusal, and Matheus Tavares Frigo. MBP: Not just an IoT platform. In *2020 IEEE International Conference on Pervasive Computing and Communications Workshops (PerCom Workshops)*, 1–3, Austin, TX. IEEE, 2020.
13. Prajit Kumar Das et al. Context-sensitive policy based security in internet of things. In *2016 IEEE International Conference on Smart Computing (SMARTCOMP)*, pp. 1–6, St Louis, MO. IEEE, 2016.
14. Ramon dos Reis Fontes et al. Mininet-WiFi: A platform for hybrid physical-virtual software-defined wireless networking research. In *Proceedings of the 2016 Conference on ACM SIGCOMM 2016 Conference*, pp. 607–608, Florianópolis. ACM, 2016.
15. Jayavardhana Gubbi et al. Internet of things: A vision, architectural elements, and future directions. *Future Generation Computer Systems*, 29–27:1645–1660, 2013.
16. Chris Hickman. How x.509 certificates were involved in solarwinds attack — keyfactor. `https://securityboulevard.com/2020/12/how-x-509-certificates-were-involved-in-solarwinds-attack-keyfactor/`.
17. Pascal Hirmer, Uwe Breitenbücher, Ana Cristina Franco da Silva, Kálmán Képes, Bernhard Mitschang, and Matthias Wieland. Automating the provisioning and configuration of devices in the internet of things. *Complex Systems Informatics and Modeling Quarterly*, 9:28–43, 2016.
18. James Hong et al. Demo: Building comprehensible access control for the internet of things using beetle. In *Proceedings of the 14th Annual International Conference on Mobile Systems, Applications, and Services Companion, , MobiSys '16 Companion*, pp. 102–102, Singapore. ACM, 2016.
19. Sean Dieter Tebje Kelly et al. Towards the implementation of IoT for environmental condition monitoring in homes. *IEEE Sensors Journal*, 13(10):3846–3853, 2013.
20. Florian Kohnhäuser, David Meier, Florian Patzer, and Sören Finster. On the security of IIoT deployments: An investigation of secure provisioning solutions for OPC UA. *IEEE Access*, 9:99299–99311, 2021.
21. Dave Kosiur. *Understanding Policy-Based Networking*, vol. 20. John Wiley & Sons, Hoboken, NJ, 2001.
22. Sungmoon Kwon, Jaehan Jeong, and Taeshik Shon. Toward security enhanced provisioning in industrial IoT systems. *Sensors*, 18(12):4372, 2018.
23. Amit A. Levy et al. Beetle: Flexible communication for bluetooth low energy. In *Proceedings of the 14th Annual International Conference on Mobile Systems, Applications, and Services, MobiSys '16*, pp. 111–122, New York, NY. ACM, 2016.
24. Chieh-Jan Mike Liang et al. SIFT: Building an internet of safe things. In *Proceedings of the 14th International Conference on Information Processing in Sensor Networks*, pp. 298–309, Seattle, WA. ACM, 2015.
25. Minzhao Lyu et al. Quantifying the reflective ddos attack capability of household IoT devices. In *Proceedings of the 10th ACM Conference on Security and Privacy in Wireless and Mobile Networks*, pp. 46–51. ACM, 2017.
26. Alex Mavromatis, Aloizio Pereira Da Silva, Koteswararao Kondepu, Dimitrios Gkounis, Reza Nejabati, and Dimitra Simeonidou. A software defined device provisioning framework facilitating scalability in internet of things. In *2018 IEEE 5G World Forum (5GWF)*, pp. 446–451, Santa Clara, CA. IEEE, 2018.

27. Carolina V. L. Mendoza et al. Defense for selective attacks in the IoT with a distributed trust management scheme. In *2016 IEEE International Symposium on Consumer Electronics (ISCE)*, pp. 53–54, Sao Paulo. IEEE, 2016.
28. Markus Miettinen et al. IoT sentinel: Automated device-type identification for security enforcement in IoT. In *2017 IEEE 37th International Conference on Distributed Computing Systems (ICDCS)*, pp. 2177–2184, Atlanta, GA. IEEE, 2017.
29. Stefan Nastic, Hong-Linh Truong, and Schahram Dustdar. A middleware infrastructure for utility-based provisioning of IoT cloud systems. In *2016 IEEE/ACM Symposium on Edge Computing (SEC)*, pp. 28–40, Washington, DC. IEEE, 2016.
30. Antonio L. Maia Neto et al. AoT: Authentication and access control for the entire IoT device life-cycle. In *Proceedings of the 14th ACM Conference on Embedded Network Sensor Systems CD-ROM*, pp. 1–15, Stanford, CA. ACM, 2016.
31. Yin Minn Pa Pa et al. IoTPOT: Analysing the rise of IoT compromises. *EMU*, 9:1, 2015.
32. Luis Alberto B. Pacheco et al. Evaluation of distributed denial of service threat in the internet of things. In *2016 IEEE 15th International Symposium on Network Computing and Applications (NCA)*, pp. 89–92, Cambridge, Boston, MA. IEEE, 2016.
33. Pavan Pongle et al. A survey: Attacks on RPL and 6LoWPAN in IoT. In *2015 International Conference on Pervasive Computing (ICPC)*, pp. 1–6, Pune. IEEE, 2015.
34. Jim Sermersheim. Lightweight directory access protocol (LDAP): The protocol. Technical report, 2006.
35. Susan J. Shepard. Policy-based networks: Hype and hope. *It Professional*, 2(1):12–16, 2000.
36. Dylan Smyth et al. Exploiting pitfalls in software-defined networking implementation. In *2016 International Conference on Cyber Security and Protection of Digital Services*, pp. 1–8. IEEE, 2016.
37. Sina Sontowski, Maanak Gupta, Sai Sree Laya Chukkapalli, Mahmoud Abdelsalam, Sudip Mittal, Anupam Joshi, and Ravi Sandhu. Cyber attacks on smart farming infrastructure. In *2020 IEEE 6th International Conference on Collaboration and Internet Computing (CIC)*, pp. 135–143, Atlanta, GA. IEEE, 2020.
38. Patrícia R Sousa, Joao S Resende, Rolando Martins, and Luís Antunes. Secure provisioning for achieving end-to-end secure communications. In *International Conference on Ad-Hoc Networks and Wireless*, pp. 498–507, Luxembourg. Springer, 2019.
39. Ioannis Stellios, Panayiotis Kotzanikolaou, and Mihalis Psarakis. Advanced persistent threats and zero-day exploits in industrial internet of things. In Cristina Alcaraz (ed.), *Security and Privacy Trends in the Industrial Internet of Things*, pp. 47–68. Springer, Berlin, 2019.
40. Mark Stevens, W. Weiss, H. Mahon, B. Moore, J. Strassner, G. Waters, A. Westerinen, and J. Wheeler. Policy framework. *Work in Progress*, 1999.
41. Andrés Villa-Henriksen, Gareth T. C. Edwards, Liisa A. Pesonen, Ole Green, and Claus Aage Grn Srensen. Internet of things in arable farming: Implementation, applications, challenges and potential. *Biosystems Engineering*, 191:60–84, 2020.
42. Bogdan Visan et al. Vulnerabilities in hub architecture IoT devices. In *2017 14th IEEE Annual Consumer Communications & Networking Conference (CCNC),*, pp. 83–88, Las Vegas, NV. IEEE, 2017.
43. Michael Vögler, Johannes Schleicher, Christian Inzinger, Stefan Nastic, Sanjin Sehic, and Schahram Dustdar. Leonore–large-scale provisioning of resource-constrained IoT deployments. In *2015 IEEE Symposium on Service-Oriented System Engineering*, pp. 78–87, San Francisco Bay, CA. IEEE, 2015.

44. Dong Wang, Sooyong Lee, Yongsheng Zhu, and Yuguang Li. A zero human-intervention provisioning for industrial IoT devices. In *2017 IEEE International Conference on Industrial Technology (ICIT)*, pp. 1271–1276, Toronto. IEEE, 2017.
45. Attila Altay Yavuz. ETA: Efficient and tiny and authentication for heterogeneous wireless systems. In *Proceedings of the Sixth ACM Conference on Security and Privacy in Wireless and Mobile Networks*, pp. 67–72, Budapest. ACM, 2013.
46. Ilker Yavuz and Berna Ors. End-to-end secure IoT node provisioning. *Journal of Communications*, 16(8):341–346, 2021.
47. Zhao Yongchi et al. Analysis of ARP spoof attack. *Programmable Controller & Factory Automation*, 2:017, 2005.
48. Ce Zhang and Azim Eskandarian. A survey and tutorial of EEG-based brain monitoring for driver state analysis. *arXiv preprint arXiv:2008.11226*, 2020.
49. Pei Yun Zhang, Yu Tong Chen, Meng Chu Zhou, Ge Xu, Wen Jun Huang, Yusuf Al-Turki, and Abdullah Abusorrah. A fault-tolerant model for performance optimization of a fog computing system. *IEEE Internet of Things Journal*, 9(3):1725–1736, 2021.

5 A Privacy-Sensitive, Situation-Aware Description Model for IoT

Zakaria Maamar
University of Doha for Science and Technology

Amel Benna
Research Center for Scientific and Technical Information (CERIST)

Noura Faci
University of Lyon, UCBL, CNRS, INSA Lyon, LIRIS

Fadwa Yahya
Prince Sattam bin Abdulaziz University, University of Sfax

Nacereddine Sitouah
Polytechnic University of Milan

Wassim Benadjel
University of Science and Technology Houari Boumédiene

5.1 INTRODUCTION

According to the International Data Corporation (IDC), *"IoT spending will increase by a compound annual growth rate (CAGR) of 13.6% from 2017 to 2022, reaching $1.2 trillion within the next four years"*.[1] In line with IDC figures, *"A research published by Transforma Insights revealed that the number of active IoT devices globally is expected to grow from 7.6 billion in 2019 to 24.1 billion in 2030"*.[2] Despite the popularity of the Internet of Things (IoT) among the information and communication technologies (ICT) community, different challenges still persist undermining this popularity among other things. Our personal list of challenges includes diversity and multiplicity of things' development and communication technologies [16], users' reluctance, and sometimes rejection because of privacy invasion that things cause [24,35], limited IoT-platform interoperability [11], lack of killer applications that would justify the existence of things [30], lack of an IoT-oriented software engineering discipline that would guide the analysis, design, and develop-

DOI: 10.1201/9781003199410-5

ment of things [21,49], and, finally, passive nature of things that primarily act as data suppliers (with some actuating capabilities) [18,37]. For more challenges, readers could refer to Refs. [22,39].

Over the years, we tackled some of the challenges above through initiatives related to federation-of-things [31], cognitive things [32], process-of-things [33], and quality-of-things model [40]. In this chapter, we tackle the particular challenge of *privacy invasion* that becomes acute when compounded with security concerns [1]; connected devices/things can give hackers and cyber-criminals more entry points.[3] When it comes to privacy, IoT exemplifies Mark Weiser's definition of ubiquitous computing when he states in 1999 that "*The most profound technologies are those that disappear. They weave themselves into the fabric of everyday life until they are indistinguishable from it*" [45]. Whether visible or invisible, today's things like thermal cameras, motion sensors, and wrist watches collect a plethora of details about persons' (and even about other things) habits, practices, preferences, and choices. For Ziegeldorf et al., "*the increasingly invisible, dense and pervasive collection, processing and dissemination of data in the midst of people's private lives gives rise to serious privacy concerns. Ignorance of those issues can have undesired consequences*" [24]. While the collection, processing, and distribution of details about users could be subject to strict approvals, seeking approval for every single detail and continuously would become cumbersome and over time inefficient due, sometimes, to people's "easy-going" nature. In addition, the lack of a systematic method for integrating privacy into IoT applications' development life cycle is exacerbating privacy as per Perera et al. who suggest privacy-by-design framework [14]. Completely different from such a framework that suggests how to minimize data acquisition and encrypt data storage that are independent from things, we directly act upon things' descriptions from a contextual perspective resulting into a better control of what things can do *versus* cannot do, so that users' privacy is preserved. To achieve this control, we define the necessary metamodels that would describe things and their operations, capture privacy concerns, relate these concerns to user situations, and allow/disallow these operations according to these situations. Based on these metamodels, we generate a privacy-sensitive, situation-aware thing description metamodel in compliance with model-driven architecture (MDA, [41]) that is known for its capability of supporting metamodel/model separation and transformation. Our objectives are, first, to inject privacy-driven contextual elements into a thing's description and, second, to transform the injected thing's description into a specific implementation that would accommodate the thing's technical specification. For illustration purposes, the Web-of-Things (WoT) Thing Description (TD) is used to describe things and will be adjusted to ensure the awareness of things to privacy concerns.

To achieve a context-driven invasion of privacy (in the sense of a controlled invasion that would comply with the privacy definition of Westin, "*the right to select what personal information about me is known to what people*" [46]), we propose an MDA compliant approach of three stages. In stage 1, we define SituationPrivacy metamodel to abstract the key concepts of privacy, situation, and operation. In stage 2, we use SituationPrivacy to transform WoT TD metamodel into a privacy-sensitive, situation-aware WoT TD metamodel. Finally, in stage 3, we automatically generate a privacy-sensitive, situation-aware WoT TD model using a set of transformation rules that we apply to a case study about elderly people in a care center. Our contributions include, but not limited to, (i) MDA weaving into IoT to preserve privacy, (ii) development of necessary metamodels and models in compliance with MDA, (iii) enrichment of WoT TD to develop things that are sensitive to privacy and aware of situations, and (iv) a system demonstrating the technical doability of MDA weaving into IoT. While WoT TD illustrates our approach, the solutions put forward in this chapter could be integrated into other thing specifications like IoT-DDL [28], CoRE-TD [17], WoT-AD [29], and O-DF[4] (though these specifications' descriptions remain restricted to research projects). The rest of this chapter is organized as follows. Section 5.2 discusses privacy and MDA in IoT and, then, presents a case study. Sections 5.3 and 5.4 are dedicated to the approach for designing, developing, and demonstrating privacy-sensitive, situation-aware things. Section 5.5 concludes the chapter and identifies future research directions.

5.2 BACKGROUND

This section discusses privacy in IoT, examines the synergy between MDA and IoT, provides an overview of WoT TD, and, finally, suggests a case study that will be used throughout the chapter to illustrate how things would end up sensitive to privacy and aware of situations.

5.2.1 PRIVACY IN IoT IN-BRIEF

Although privacy is a recurrent concern in the ICT community, it becomes acute in the IoT since things are "invited" to be part of our lives. According to Orman [38], *"The things in the Internet of Things (IoT) can get personal. They can be in your home, your car, and your body. They can make your living and working space smart, and they can be* ***dangerous*** *to your health, safety, and liberty ... Is our future a brave new world or a dystopian nightmare? Who decides?"* On top of users' approvals to let things be part of their lives [44] and the inappropriateness of existing privacy-preserving techniques like k-anonymity and secure multi-party computation for industrial IoT [12], many device manufacturers exacerbate the privacy concern when they suspend or stop upgrading their devices in response to specific threats that would impact users' privacy [23]. This sporadic and discontinued upgrade gives intruders sufficient time to crack security protocols and access private data.

Compounded to device manufacturers' passiveness, organizations including governments are sometimes unable to enforce their own privacy policies in response to specific regulatory demands like disclosing airline passenger details [10,13]. It becomes quite impossible to predict all possible privacy violation scenarios despite the goodwill of organizations and technical solutions to preserve users' privacy. Briefly, these solutions could either be split into access control and blockchain-based [48] or refer to intrusion detection systems [20]. The former rely on encryption where either access policies or associated attributes are kept secret, and the latter guarantee integrity and privacy of sensitive data where only authenticated recipients decrypt authorized transactional details.

5.2.2 DEFINITIONS

5.2.2.1 Internet of Things

A comprehensive guide about IoT has been released by the DZone group in 2017 [18]. The guide covers various aspects relevant for IoT such as privacy, big data, monitoring, context, and architecture. In Ref. [4], Barnaghi and Sheth provide a good overview of IoT requirements and challenges. Requirements include quality, latency, trust, availability, reliability, and continuity that should impact efficient access and use of IoT data and services. And, challenges result from today's IoT ecosystems featuring billions of dynamic things and thus making existing search, discovery, and access techniques and solutions inappropriate for IoT data and services. In Ref. [2], Abdmeziem et al. discuss IoT characteristics and enabling technologies. Characteristics include distribution, interoperability, scalability, resource scarcity, and security, while enabling technologies include sensing, communication, and actuating. These technologies are mapped onto a three-layer IoT architecture consisting of perception, network, and application, respectively.

5.2.2.2 Model-Driven Architecture

Different studies discuss MDA for system design and development [3,9,19,47]. The complete design model of an application comprises multiple models offering each a different viewpoint of the application. For instance, UML offers models ranging from use-case and class diagrams to deployment diagrams, which are used depending on the objectives to achieve. In MDA, the emphasis is on designing detailed models and platform independent specifications. For Backx, MDA addresses the migration problems between computing platforms, which are prone to errors and time consuming [3]. By having models that are language and platform independent, same models can be applied to different scenarios. To ensure the acceptance of MDA, the following models are produced: Computation independent model (CIM) for business requirements (or business model), platform independent model (PIM) for system functionalities, and platform-specific model (PSM) for platform-specific details. Initially, the design work starts with preparing a CIM that will be mapped onto a PIM. This one is the starting point of deriving a PSM according to the characteristics of the target platform using specific Model-2-Model (M2M) transformation rules. Finally, a tool uses the PSM to generate the complete code of the future system.

5.2.3 WHEN MDA MEETS IoT

In Ref. [26], Im et al. adopt MDA to design and develop an IoT mashup-as-a-service (MaaS). The authors expose things as services so they can tap into existing solutions and technologies for service mashup like those discussed in Refs. [8,34]. IoT mashup consists of composing three models referred to as thing model, software model, and computational-resource model. Along with these three models, Im et al. describe things using first, a PIM to address heterogeneity across things and second, a PSM once a thing implementation technology is agreed upon.

In Ref. [42], Sosa-Reyn et al. present a method based on model-driven engineering (MDE) to design and develop IoT applications. They characterize things, whether physical or virtual, with identities, physical attributes, and virtual personalities that have pervasive sensing, detection, actuation, and computational capabilities. Thanks to MDE's abstraction and granularity levels, different models are designed and transformed until the necessary code of future IoT applications is generated. In conjunction with MDE, Sosa-Reyn et al. adopt service-oriented architecture (SOA) to propose an architecture of four layers, namely, object, network, service, and application. Sosa-Reyn et al.'s MDE-based method includes four phases, namely, analysis of business requirements (using UML's use-case and activity diagrams in compliance with PIM principles), definition of business logic (using BPMN language in compliance with PIM principles again), design of integrated services solution, and generation of technological solution (using a particular implementation technology in compliance with PSM principles).

In Ref. [43], Thang Nguyen et al. present a Framework for sensor application development (FRASAD) by adopting MDA principles to address the complexity of developing such applications. FRASAD adopts a node-centric software architecture and a rule-based programming model allowing both to obtain code applications through a set of automatic model transformation steps. FRASAD's architecture includes five layers, namely, application, operating system abstraction, operating system, hardware abstraction, and hardware, allowing to transform an application model into an equivalent final platform specific-application. Thang Nguyen et al. associate FRASAD with a graphical user interface to describe IoT applications using a domain-specific language.

In Ref. [15], Ciccozzi and Spalazzese introduce MDE4IoT standing for model-driven engineering for internet of things. MDE4IoT allows the modeling of things along with supporting their self-adaptation in the context of emergent configurations that capture the temporary cooperation and connection between things. Temporary because things might become unavailable due to physical mobility or insufficiency battery level or even battery problems. Should this be the case, run-time management of emergent configurations would be necessary by ensuring that things' software functionalities would be abstracted away from platform-specific details. MDE4IoT helps achieve this abstraction, so that executable artifacts are generated.

In Ref. [6], Berrouyne et al. use the acronym AAA for anything, anytime, and anywhere to define IoT and discuss the challenges of achieving interoperability in IoT. Although things are by nature heterogeneous, they are expected to work together to provide value-added services to users. Berrouyne et al. suggest MDE to address thing heterogeneity that could happen at different levels such as functionality, programming, and communication. Atlas transformation language (ATL, [27])-based M2M transformation was adopted allowing, for instance, to adapt models of things' behaviors according to models of networks allowing a certain form of compatibility between things and networks.

Although the afore-mentioned works offer a sample of mixing MDA and IoT together, there is a gap in both examining privacy in IoT and taping into MDA to address this gap. We embrace MDA's principles and mechanisms to ensure that things' descriptions are sensitive to users' privacy concerns and accommodate measures that address these concerns.

5.2.4 WoT TD IN-BRIEF

As stated in Section 5.1, we use WoT TD[5] for describing things. WoT TD represents a central building block in the W3C Web of Things (WoT) and can be considered as a thing's entry point. A thing is an abstraction of either a physical or a virtual entity whose metadata and interfaces are described by a WoT TD. The WoT TD conceptual model is built upon four parts referred to as Vocabulary having each a namespace: (i) *core TD* Vocabulary defines the interaction model, (ii) *Data Schema* Vocabulary describes a common subset of the terms defined in a JSON Schema, (iii) *WoT Security* Vocabulary identifies the configuration of the security mechanisms, and (iv) *Hypermedia Controls* Vocabulary provides a representation for Web *links* and Web *forms* that a Thing exposes to potential end-users.

Figure 5.1 is an excerpt of WoT TD metamodel prior to any adjustment that would address the particular concern of privacy. We focus on WoT TD's elements that correspond to the main meta-classes of *core TD* Vocabulary along with the meta-classes (shown in gray) of the three other independent vocabularies of WoT TD that are *DataSchema* meta-class for *Data Schema* Vocabulary, *SecurityScheme* meta-class for *WoT Security* Vocabulary, and *Form & Link* meta-classes for *Hypermedia Controls* Vocabulary. Still in Figure 5.1, thing offers three choices of InteractionAffordance that show how end-users and/or peers could interact with a thing: *Properties* of type PropertyAffordance, *Actions* of type ActionAffordance, and *Events* of type EventAffordance. *Properties* allows to sense and control parameters, *Actions* corresponds to a thing's operations, and *Events* allows to asynchronously push communications such as notifications, discrete events, and streams of values to receivers.

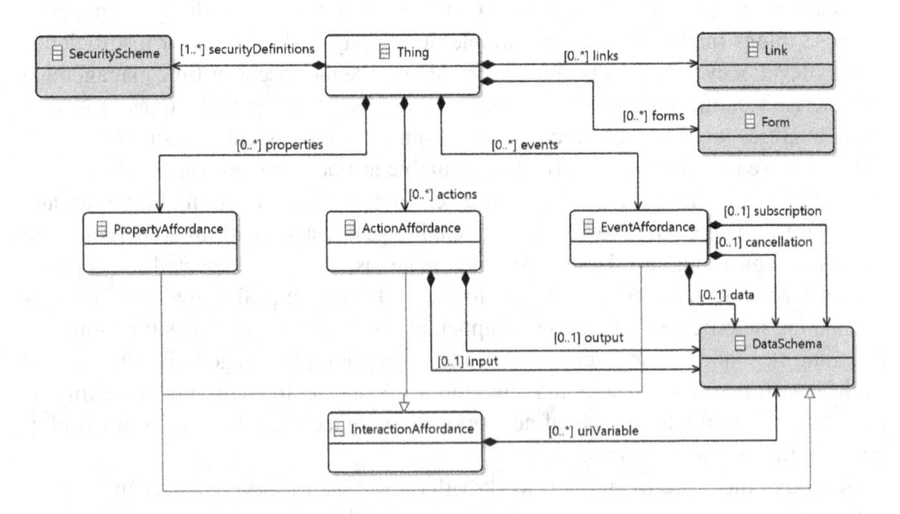

Figure 5.1 Excerpt of the initial WoT TD metamodel diagram.

Listing 5.1 Excerpt of WoT TD Specification of Lamp

```
1{
2    "@context": "https://www.w3.org/2019/wot/td/v1",
3    "id": "urn:dev:ops:32473-WoTLamp-1234",
4    "title": "MyLampThing",
5    "securityDefinitions": {
6        "basic_sc": {"scheme": "basic", "in":"header"}},
7    "security": ["basic_sc"],
8    "properties": {
9        "status": {
10           "type": "string",
11           "forms": [{"href": "https://mylamp.example.com/status"}]}
12       },
13   "actions": {
14       "toggle": {
15           "forms": [{"href": "https://mylamp.example.com/toggle"}]}
16       },
17   "events": {
18       "overheating": {
19           "data": {"type": "string"},
20           "forms": [{
21               "href": "https://mylamp.example.com/oh",
22               "subprotocol": "longpoll"}]}
23       }
24}
```

Listing 5.1 shows a WoT TD instance of a lamp Thing referred to as MyLampThing. Thanks to this description, we know that there exists one property affordance with the title *status* (line 9), an action affordance is specified to *toggle* (line 14) the switch status, and an Event affordance known as *overheating* (line 18) that enables a mechanism for asynchronous messages to be sent by a thing. The listing also specifies a basic security scheme requiring username and password for access (line 7).

5.2.5 CASE STUDY

With the latest ICT advances, many personal details about people are collected from different sources and afterward shared without seriously asking who will receive and process these details and for what purposes like understanding people's habits, movements, and even feelings [50]. Our case study sheds light on privacy in elderly care centers. Many studies confirm that population aging is a dominant global demographic trend of the 21st century.[6] Despite the benefits of monitoring the patients of these centers, a good number of privacy concerns can be identified.

According to Ref. [25], the five most important smart tech-devices for senior safety[7] include smart home hub, smart home sensors, smart lights, smart medication dispensers (SMD), and smart stove shutoff. Let us consider the situation where a center's patients get together every Thursday's afternoon to watch movies in the living room. Each patient is expected to have her own SMD when she is not in her room. Prior to the viewing session, the smart TV synchronizes with all attendees' dispensers so that reminders are displayed on the screen and, if required, alerts are sent to the medical staff. While this synchronization is mandatory, patients' and medicines' details could be used by third parties to develop targeted healthcare awareness campaigns for these patients making them subscribe to new programs. In Ref. [5],

Table 5.1
High-Level Specification of Some Smart Devices in the Case Study

Type	Operation	Description
Smart Medication Dispenser	`change`	Modify the intake frequency per type of pill
	`Configure`	Select the container per type of pill
	`Dispense`	Release a pill in a container and blink lights
	`Display`	Notify availability of pills on the TV screen
	`Refill`	Request for more pills
Smart TV	`Display`	Enable programs for viewing
	`Record`	Tape ongoing programs
	`Trigger`	Initiate voice control
	`Share`	Send details to the cable TV company

Belhajjame et al. shed light on cases where combining medicines' names could be used to infer some diseases and hence insisted on data privacy.

To mitigate privacy concerns, the SMDs, smart TV, and other devices could automatically have their specifications (Table 5.1) adjusted according to some details like living room, Thursday's afternoon, and medicines' dosages in the afternoon. How to ensure that the dispensers' reminders displayed on the smart TV will not be relayed to the cable TV company? Which operations of the SMD should be temporarily disabled without impacting the viewing experience (e.g., TV's subtitling remains enabled) nor risking the safety of patients (e.g., SMD's alarm remains enabled too)? And, how to ensure that pairing SMDs with patients' mobile phones will not expose their personal contacts to these SMDs' vendors? There are some questions that we raise and address in the rest of this chapter.

5.3 PRIVACY-SENSITIVE AND SITUATION-AWARE THING DESCRIPTION

This section discusses how our approach acts upon the descriptions of things, so that, things become sensitive to privacy and aware of situations/contexts. After an overview of the approach's three steps, the remaining sections detail each step in terms of objectives to achieve, actions to perform, and techniques to adopt.

5.3.1 OVERVIEW

Figure 5.2 illustrates the levels and transformations allowing to inject situation-based privacy details into things' descriptions like WoT TD. This injection happens in compliance with, first, the object management group (OMG)'s guidelines that separate conceptual and technical details of a system's operations [36] and, second, the MDA's guidelines for model transformations [7]. First, the levels in this figure host models, metamodels, and meta-metamodels and ensure the instantiation of meta-metamodels into metamodels and then, metamodels into models. Second,

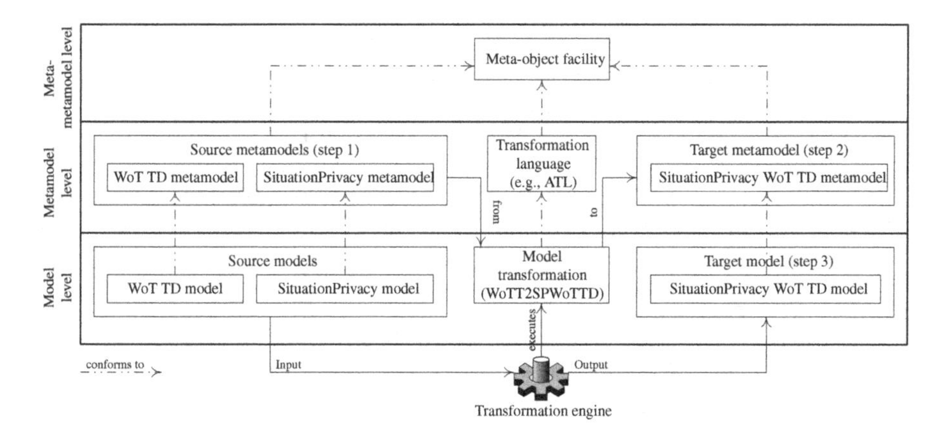

Figure 5.2 Levels and transformations for black developing privacy-sensitive, situation-aware thing descriptions (adapted from [7]).

the transformations convert privacy-insensitive, situation-unaware thing descriptions into privacy-sensitive, situation-aware thing descriptions. To accommodate all these models and transformations, our approach runs through three steps:

- Step 1 is manually completed and taps into the content of Section 5.2.1 along with our own understanding of the concepts revolving around privacy like thing, situation, and operation. The objective of this step is to define the necessary constructs of the future SituationPrivacy metamodel that will fall into the *source metamodels* box in Figure 5.2.
- Step 2 is manually completed as well and examines potential connections/overlaps between the constructs of SituationPrivacy obtained in Step 1 and the constructs of WoT TD metamodels. The objective of this Step 2 is to define the necessary constructs of the future SituationPrivacyWoTTD metamodel that will fall into the *target metamodel* box in Figure 5.2.
- Step 3 is automatically completed requiring the development of rules in, for instance, ATL [27] to transform the source metamodels' constructs defined in Step 1 into the target metamodel's constructs defined in Step 2. The objective of Step 3 is to implement these rules through an engine in order to define the necessary constructs of the future SituationPrivacyWoTTD model that will fall into the *target model* box in Figure 5.2.

5.3.2 STEP 1: SITUATIONPRIVACY METAMODEL DEFINITION

This step develops a dedicated metamodel, SituationPrivacy, whose constructs would capture users' privacy concerns according to the situations in which these concerns could arise. This metamodel shown in Figure 5.3 consists of 7 meta-classes, SituationPrivacyThing, Situation, Privacy, Operation, PrivacyRequirement, SituationProperties, and PropertyComposition, and several meta-relations between these

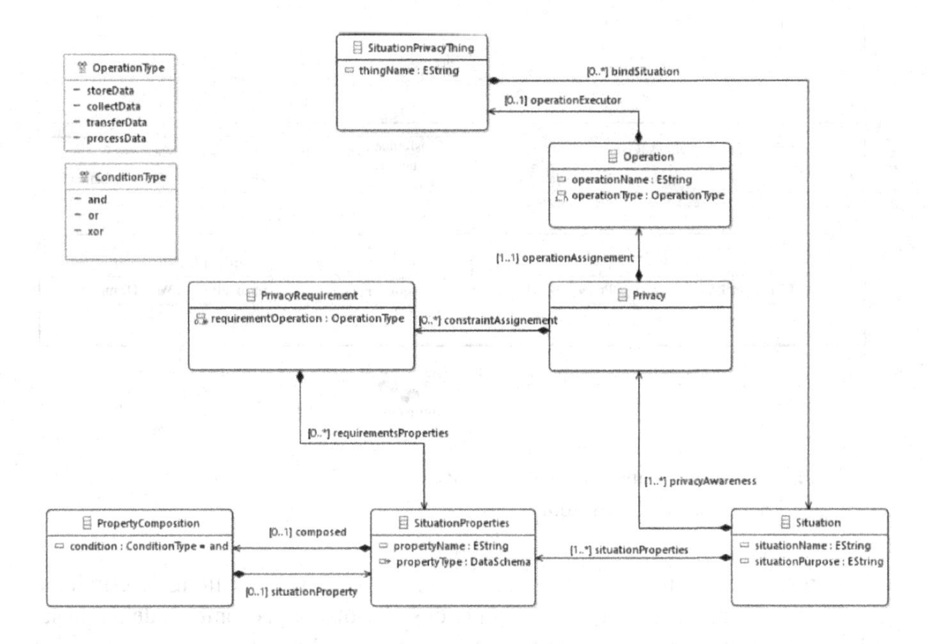

Figure 5.3 SituationPrivacy metamodel diagram.

meta-classes. First, SituationPrivacyThing meta-class abstracts concrete things like SMD and smart TV that reside in a user's cyber-physical space. This space is represented with a meta-class that is situation. At runtime, things participate in real situations so that they track changes in these situations' details such as time, location, and preferences. From a meta-modeling perspective, we abstract these details with SituationProperties and PropertyComposition meta-classes. Second, the operations that a thing uses to manipulate details about situations are abstracted with operation meta-class. The four possible operations that are deemed relevant for examining privacy in IoT are *collect*, *process*, *store*, and *transfer* and are associated with OperationType meta-attribute. Third, during detail manipulation, privacy concerns could arise like collecting details without approvals and transferring details to unauthorized third parties. We map privacy concerns onto Privacy meta-class that consists of requirements and actions represented with PrivacyRequirement and Operation meta-classes, respectively. The former meta-class refers to some operation types (defined through OperationType enumeration) that are associated with some details about a situation and the thing that should satisfy a particular requirement constraining the manipulation of these details. Back to the elderly care center (Table 5.1), an example of requirement would be to disable the operation of sharing (corresponding to an instance of requirementOperation meta-attribute with transferData as a data type) medicines' names and dosages values, should the current situation indicate the living room, during week end, and the smart TV executing the display operation.

5.3.3 STEP 2: SITUATIONPRIVACYWoTTD METAMODEL DEFINITION

This step makes things sensitive to privacy and aware of situations. To this end, we examine how some meta-classes in SituationPrivacy metamodel of Figure 5.3 could connect/overlap with other meta-classes in WoT TD metamodel of Figure 5.1. The result is SituationPrivacyWoTTD metamodel as per Figure 5.4 and is made possible because WoT TD can accept contextual knowledge from other namespaces using *TD context*. This allows to enrich TD instances with additional semantics that could be domain specific and to import additional schemes like protocol bindings and new security, if deemed necessary.

In Figure 5.4, ActionAffordance meta-class, originating from WoT TD metamodel, is a cornerstone in SituationPrivacyWoTTD metamodel by connecting to Operation and PrivacyRequirement meta-classes, both originating from SituationPrivacy metamodel. While ActionAffordance:Operation connection sets the actions that would empower a thing with respect to a user situation, ActionAffordance:PrivacyRequirement connection permits to either enable or disable these actions according to this user situation's requirements. More details about enable/disable are given in Section 5.3.4. Finally, in Figure 5.4, Thing meta-class, originating from WoT TD metamodel, inherits from SituationPrivacyThing, originating from SituationPrivacy metamodel, so that the binding between Thing and Situation meta-classes happens. Finally, PropertyAffordance meta-class, originating from WoT TD metamodel, inherits from SituationProperties meta-class, originating from SituationPrivacy metamodel, to ensure that some details reported in Situation meta-class, originating from SituationPrivacy metamodel, remain private.

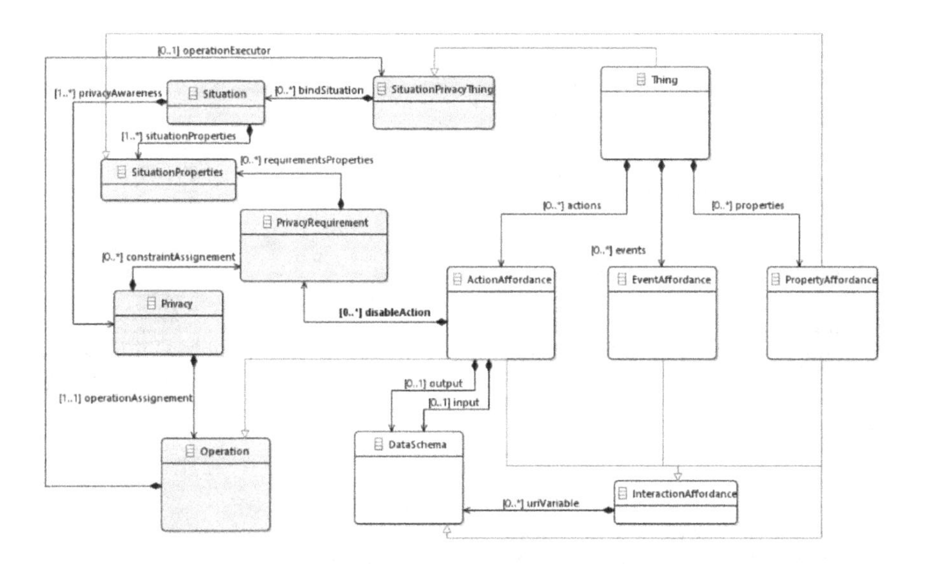

Figure 5.4 Excerpt of SituationPrivacyWoTTD metamodel diagram.

5.3.4 STEP 3: SITUATIONPRIVACYWoTTD MODEL GENERATION

To automatically generate SituationPrivacyWoTTD models that would be conform to SituationPrivacyWoTTD metamodel, we, as per Figure 5.2, defined model transformation and then developed a set of ATL rules that are submitted to a transformation engine.

In Figure 5.5's right panel, labeled as WoTT2SPWoTTD and corresponding to model transformation, a set of rules are listed. *Thing2Thing* converts each element of Thing meta-class in WoT TD metamodel into an element of Thing meta-class in SituationPrivacyWoTTD metamodel. And, *Action2Action* detailed in Listing 5.2 integrates privacy into action after converting each element of ActionAffordance meta-class in WoT TD metamodel (lines 6-16) into an element of ActionAffordance meta-class in SituationPrivacyWoTTD metamodel. As a result, a new disableAction element (lines 17–19) links ActionAffordance to Privacy (originating from SituationPrivacy metamodel) in SituationPrivacyWoTTD metamodel. This new element indicates that an operation will be disabled because of a particular situation where the privacy of some details needs to be maintained. More technical details about model transformation are given in the next section.

5.4 IMPLEMENTATION

This section discusses the efforts that were put into verifying the technical doability of our MDA approach for making things sensitive to privacy and aware of situations. The section starts with model transformation and then presents how the case study was simulated and evaluated.

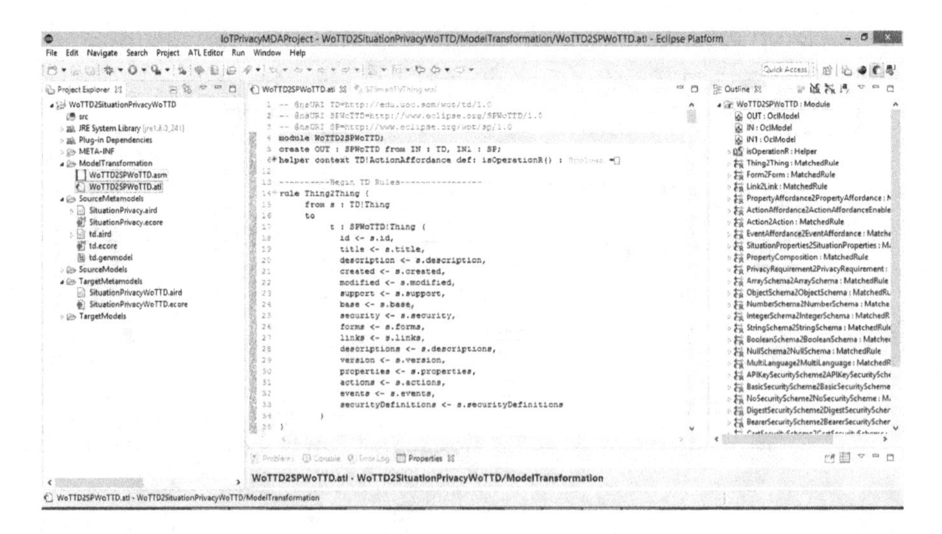

Figure 5.5 Specification of model transformation.

Listing 5.2 ATL Definition of *Action2Action* Rule

```
1 rule Action2Action {
2   from
3       s : TD!ActionAffordance(s.oclIsTypeOf(TD!ActionAffordance))
4   to
5       t : SPWoTTD!ActionAffordance (
6           id <- s.id,
7           ititle <- s.ititle,
8           ititles <- s.ititles,
9           idescription <- s.idescription,
10          idescriptions <- s.idescription,
11          forms <- s.forms,
12          uriVariable <- s.uriVariable,
13          input <- s.input,
14          output <- s.output,
15          safe <- s.safe,
16          idempotent <- s.idempotent,
17          disableAction <- SP!Privacy.allInstancesFrom('IN1') ->
18          select (o|o.operationAssignement.operationName=s.id) ->
19          collect(o|o.constraintAssignement))
20 }
```

5.4.1 MODEL TRANSFORMATION

To perform WoTT2SPWoTTD model transformation, we used Eclipse 4.12 IDE after enhancement with many modeling plug-ins such as eclipse modeling framework (EMF)[8] for developing metamodels and models, ATL IDE for transforming models, and EMF client platform (ECP) for building EMF-based client applications. We built an EMF project, imported WoT TD metamodel[9] into this project, and, then, created SituationPrivacy metamodel. The latter refines WoT TD metamodel in order to create SituationPrivacyWoTTD metamodel (Figure 5.2). Based on these different metamodels, we developed an application that creates customized SituationPrivacy and WoT TD models for things. For instance, Figure 5.6 is the result of customizing the SMD's SituationPrivacy model. It describes situations where the SMD needs to keep some details private. Should the display operation be invoked in the context of a situation like living room, then medicines' names stored in a smart medication dispenser and/or phone contacts stored in a mobile phone should be kept private.

We now describe how WoTT2SPWoTTD model transformation was applied to the case study with focus on the SMD where its initial WoT TD specification is given in Appendix 1/Listing 5.4 (excluding lines 9 and 51–101). After defining the necessary WoTT2SPWoTTD model transformation, we submitted it to the ATL transformation engine that automatically generated the SMD's SituationPrivacyWoTTD model (Figure 5.7 with focus on the red frame for the SMD's privacy requirements when display is executed). This has required setting up (i) an ATL engine configuration for the WoTT2SPWoTTD model transformation; (ii) the input and output metamodels, namely, WoT TD, SituationPrivacy, and SituationPrivacyWoTTD; and finally (iii) the input models, namely, the SMD's WoT TD and SituationPrivacy models related to the SMD's WoT TD in Figure 5.6. It is worth noting that our privacy-sensitive, situation-aware SMD SituationPrivacyWoTTD is not dependent on any particular representation format like we do in Figure 5.7 using XMI. A similar representation in JSON-LD is also given in Appendix 1/Listing 5.4.

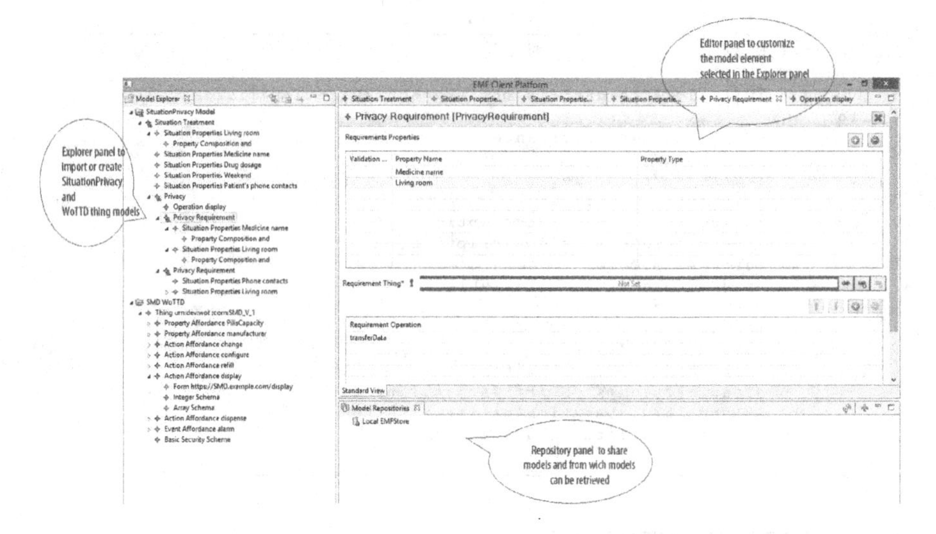

Figure 5.6 Creation of SituationPrivacy model of the smart medication dispenser.

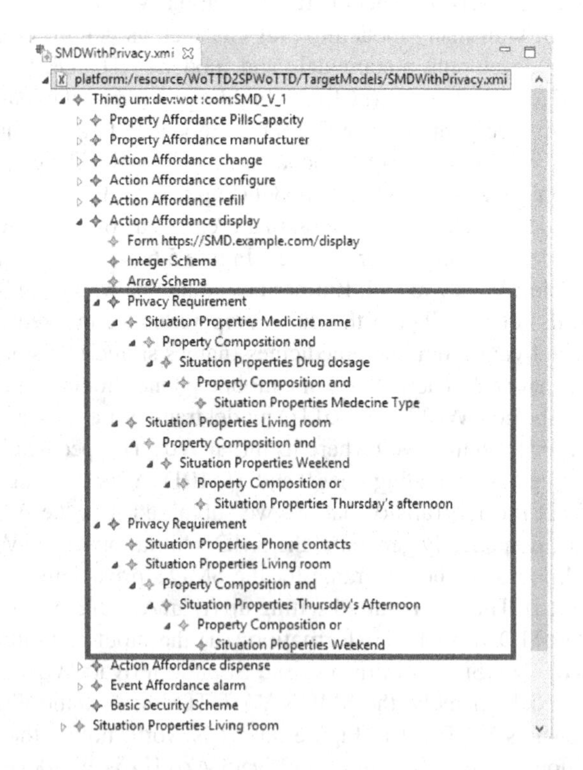

Figure 5.7 SituationPrivacyWoTTD model of the smart medical dispenser.

Listing 4.4 SMD's WoT TD Description with Some Privacy Requirements

```
1 {
2 "description": "Description of Smart Medical Dispenser",
3 "id": "urn:dev:wot:com:SMD_V_1",
4 "title": "SMDThing",
5   ...
6 "@context": [
7   "https://www.w3.org/2019/wot/td/v1",
8   {
9     "sp": "http://www.examplePrivacySchema.org/wot/sp/v1#"
10  }
11 ],
12 "securityDefinitions": {
13   ....
14 },
15 "security": ["basic"],
16 "properties": {
17       "PillsCapacity": {
18       ....
19   },
20 "actions": {
21   "change": {
22   ...
23   },
24   "configure": {
25   ....
26   },
27   "refill": {
28   ....
29   },
30   "dispense": {
31   ...
32   },
33   "display": {
34     "output": {
35       "maxItems": 30,
36       ...
37       "type": "Integer"
38     },
39     "input": {
40       ...
41       "type": "integer"
42     },
43     "forms": [
44       {
45         "href": "https://SMD.example.com/display",
46         "op": "invokeaction",
47         "contentType": "application/json"
48       }
49     ],
50     "disableAction": [
51
52       {
```

Situation Privacy namespace, referring to SituationPrivacy vocabulary

Action enhanced by privacy

new element that integrates privacy into 'display' action

```
53        "@type": "sp:PrivacyRequirement",
54        "requirementOperation": "transferData",
55         "requirementsProperties":[
56           {
57             "propertyName": "Medicine name",
58             "composed": {
59               "situationProperty":{
60                 "propertyName": "Drug dosage",
61                 "composed": {
62                   "situationProperty": {
63                     "propertyName": "Medecine Type"
64
65               ...}}}}
66           },
67           {
68             "propertyName": "Living room",
69             "composed": {
70               "situationProperty": {
71                 "propertyName": "Weekend",
72                 "composed": {
73                   "condition": "or",
74                   "situationProperty": {
75                     "propertyName": "Thursday's
76                                      afternoon"
77           ...}}}}
78           }
79         ]
80       },
81       {
82         "@type": "sp:PrivacyRequirement",
83         "requirementOperation": [
84           "collectData", "processData","transferData","storeData"
85         ],
86         "requirementsProperties": [
87           {
88             "propertyName": "Phone contacts"
89           },
90           {
91             "propertyName": "Living room",
92             "composed": {
93               "situationProperty": {
94                 "propertyName": "Thursday's Afternoon",
95                 "composed": {
96               ...}}}
97           }
98           ],
99     }]
100
101      }...
102      },
103 "events": {
104    "alarm": {
105    ...},
106    }
107 }
```

isabled operation type

personal details that needs privacy

situation where ensuring the privacy of some details is required

Privacy requirements for 'display' action

5.4.2 SIMULATION

For a complete simulation of the case study, more tools have been used, for instance, Thingweb Node-WoT,[10] Node-Red,[11] Node-red-contrib-web-of-things,[12] and Node-generator.[13] First, Thingweb Node-WoT is a WoT TD parser that creates WoT applications based on a set of IoT resources exposed as web resources. Thingweb Node-WoT uses what is called Servient that could simultaneously act as a server and client according to future WoT applications' needs and requirements. Second, the rest of tools permit to wire things together using interaction flows and visualize these flows. Briefly, Node-Red is a visual tool based on flow-based programming paradigm largely used for developing IoT applications in terms of nodes and flows. In conjunction with Node-Red, Node-generator and Node-red-contrib-web-of-things use TD to generate WoT client nodes. We carried out the development and experiments on a Lenovo Thinkpad with Core(TM) i7-8665U (4 Cores, 8 Threads, 8 MB Cache) processor, 16 GB RAM, and 64 bits Windows 10.

Based on the high-level specification of some of the case study's smart devices mentioned in Table 5.1 along with some functional requirements that we set like reminding patients about their medicines and alerting the medical staff, we used Thingweb Node-WoT to create 2 Servients associated with three TDs, namely, SMD_1, SMD_2, and smart TV (Figure 5.8). Only SMD_1's TD was made sensitive to privacy without altering its regular workflow. Whenever a patient's SMD clock matches her medicine's intake time, an action known as *dispense* is triggered setting the right quantity of pills as per the patient's prescription. This match also leads to triggering the *display* action allowing to send necessary messages to the smart TV for display.

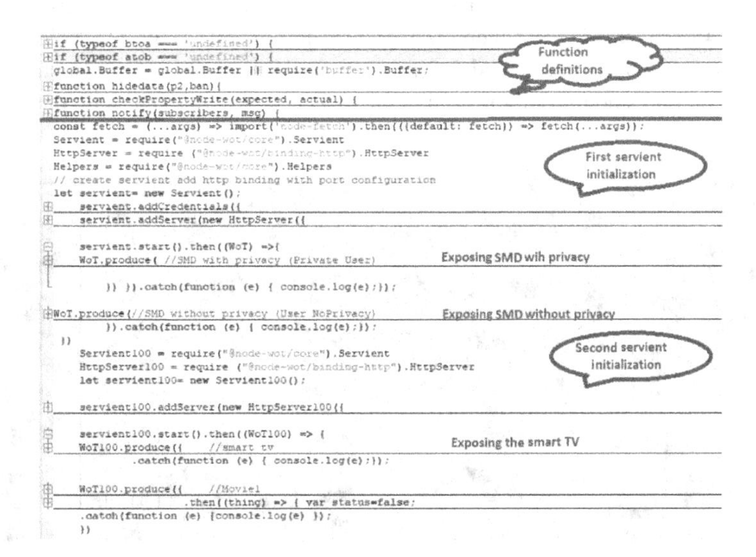

Figure 5.8 Servient definition using Thingweb Node-WoT.

By making SMD_1's *display* action sensitive to privacy thanks to *disableAction* (e.g., Appendix 1/Listing 5.4-line 51), a set of privacy requirements had to be defined (e.g., Appendix 1/Listing 5.4 lines 54-82) and associated with a JavaScript (JS) object that is conform to SituationPrivacy vocabulary. In this listing, *transferData* operation-type indicates that SMD_1 is not allowed to share details like *medicine name*, *drug dosage*, and *medicine type* with the smart TV. Contrarily, other operation-types like *collectData* and *storeData*, not shown in the listing, indicate that SMD_1 is not this time allowed to self-collect and self-store these details.

In addition to Appendix 1/Listing 5.4, we show in Figure 5.9 an excerpt of the code associated with the privacy requirements on *disableAction* element. When the *display* action is triggered in SMD_1, this latter scans the privacy requirements that would identify which details must be kept private (Figure 5.9-line 8) based on the current situation like "afternoon," "weekend," "being in the living room" (Figure 5.9-lines 9–11), and according to the operation type (Figure 5.9-lines 14–15). Afterward, SMD_1 sends the smart TV the non-private details, while the private ones are replaced with N/A (not available) value. These interactions between SMD_1 and the smart TV and other interactions with SMD_2 are captured thanks to Node-Red (Figure 5.10) and different messages (Figure 5.11).

5.4.3 EVALUATION

We proceeded with three stages targeting SMD_2 once and SMD_1 twice (Table 5.2). While 1, 5, and 10 instances of SMD_2 were under execution separately, we recorded the best and worst response times and calculated the average after 10 dispenses. Regarding SMD_1, we took into account the number of privacy requirements included in the *display* action. We started with 1 privacy requirement like in Appendix 1/Listing 5.4 (lines 53–81) and then 20 privacy requirements. Table 5.2 indicates that the average response-time related to 1 privacy requirement in SMD_1 was under one-tenth of a second, and even after running 10 SMD_1s on a single Servient, the average

```
1
2   Data_Collectable = Get.Patient_Dispense_details()
3   Data_Shareable = Get.Patient_Dispense_details()
4   For each element in $Disableaction {
5       for each element_j in element( requirementProperties ) {
6           do {
7               Get element_j(condition)_if previous exists; //(or, and, xor ..etc)
8               Get PropertyName.Type // MedecineName, DrugDosage,...etc
9               if element_j(PropertyName) is in Location_list { Verify the SMD location }
10              if element_j(PropertyName) is in Days_list  { Verify the Date }
11              if element_j(PropertyName) is in Time_list { Verify the clock }
12          } While element_j(composed) Exists
13      }
14      if element(requirementOperation) is (collectData){ HideData(Data_Collectable) }
15      if element(requirementOperation) is (storeData) { HideData(Data_Shareable) }
16  }
17
```

Figure 5.9 Excerpt of the code associated with *disableAction*.

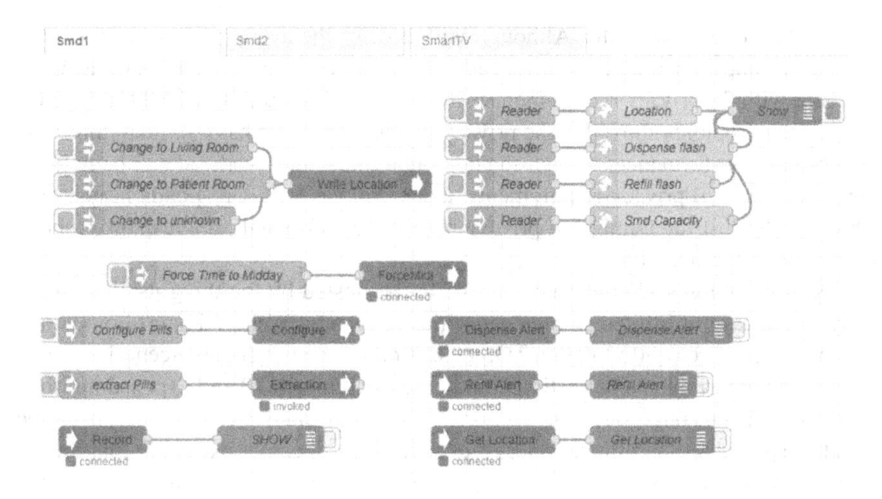

Figure 5.10 Partial representation of SMD_1's interaction flow in Node-Red.

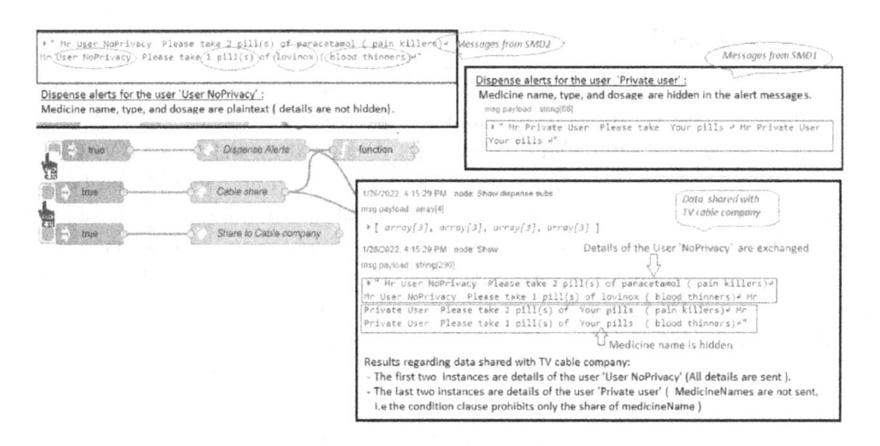

Figure 5.11 Examples of messages that SMD_1, SMD_2, and the smart TV produce.

Table 5.2
SMDs' Response Times in Milliseconds

	Number of SMD_2			Number of SMD_1 (1 privacy requirement)			Number of SMD_1 (20 privacy requirements)		
Cases	1	5	10	1	5	10	1	5	10
Best of 10	38	87	124	95	188	197	962	1,294	3,166
Worst of 10	132	255	484	189	657	1,115	1,185	6,216	19,270
Average	81	108	225	139	358	462	1,104	3,068	10,389

was slightly over two-tenths. Although with SMD_1, the average response-time surpassed 10s and in some cases, it reached 20s when running 10 SMD_1s on a single Servient with 20 privacy requirements. The results with either few SMD_1s or fewer privacy requirements show that the privacy inclusion process has a limited impact on the system's performance. In fact, the computational-time overhead can be negligible. Even with 20 privacy requirements, a stand-alone SMD_1 responds in just over a second. And, if the number of privacy requirements is small, the response time is very close to the one of SMD_2.

Because a thing's TD instance can usually be hosted by the thing itself or externally in the case of limited storage capacities or when a Web of Things-compatible legacy device is retrofitted with a TD, we examined the worst-case scenarios where the privacy requirements per action could potentially reach 20 and exposed multiple TDs (up to 20) through one Servient, only, instead of many we concluded from the results, especially when the number of exposed TDs does not exceed 5, that the response time is very reasonable (i.e., "few seconds") even when the number of privacy requirements is very high.

5.5 CONCLUSION

Prior to concluding, we discuss how the questions (at least some) we raised when presenting the case study about the elderly center are addressed. These questions are how to ensure that the SMDs' reminders displayed on the smart TV will not be relayed to the cable TV company, which operations of an SMD should be temporarily disabled without impacting the viewing experience nor risking the safety of patients, and how to ensure that pairing of SMDs with patients' mobile phones will not expose these patients' personal contacts to these SMDs′ vendors? First, we prevented the smart TV from relaying private details like patients' names by disabling the share operation in its WoT TD specification. Second, disabling the SMD' operations like refill and configure will not impact the viewing experience. Finally, the pairing will not let mobile phones access personal contacts of the SMDs' owners. All these adjustments are carried out because of the current situation of the elderly center's guests watching movies in the living room during weekends.

Despite IoT's bright side exemplified with many benefits, some persistent concerns are undermining these benefits with focus on privacy invasion that things could cause to users. This invasion's consequences could lead to identifying, tracking, and profiling users [24]. Whether invisible or visible, things collect details about persons sometimes without their knowledge/approvals. This collection is associated with operations that things execute and are part of things' descriptions. To mitigate privacy invasion, we advocated for reviewing descriptions of things in a way that operations are either enabled or disabled according to specific situational elements (context). To achieve this mitigation, we took three actions. The first action consisted of designing and developing an MDA-based approach. The second action consisted of using WoT TD to exemplify things' descriptions. Finally, the third action consisted of using an elderly care center as a case study to illustrate both the approach and the revised WoT TD. The objective is to make things sensitive to privacy concerns and aware of situational details. In compliance with MDA principles, two metamodels (SituationPrivacy and SituationPrivacyWoTTD) and a set of transformation rules in ATL were developed allowing to automatically generate SituationPrivacyWoTTD models for things like SMDs that could be the source of violating privacy. The implementation demonstrated the technical doability of our privacy-sensitive, situation-aware thing description approach where different technologies and tools were used illustrating, for instance, how a smart device's targeted actions are disabled because of specific situations and how private details are hidden because of these situations as well. Finally, although our privacy-sensitive, situation-aware thing description approach uses WoT TD, the approach remains generic and independent from any specification for thing description. In term of future work, we would like to examine the impact of enriching thing description with privacy on their discovery, consider other things' descriptions besides WoT TD, and apply SituationPrivacyWoTTD models to other case studies.

APPENDIX 1

To represent SituationPrivacyWoTTD model in JSON-LD that is WoT TD's standard representation format, two elements are needed: (i) a new namespace that would refer to SituationPrivacy vocabulary and (ii) JSON representation of SituationPrivacyWoTTD model (e.g., SMD in Figure 5.7) that would be enhanced with this namespace using WoT TD @context mechanism. To obtain these two elements, we rely on MDA principles and EMF implementation to automatically transform first, SituationPrivacy Ecore metamodel into RDF-based OWL and second, SituationPrivacyWoTTD Ecore instances into JSON. We used Ecore2OWL plugin[14] to perform the first transformation whose result is SituationPrivacy vocabulary.[15] This vocabulary extends WoT TD context knowledge as per Appendix 1/Listing 5.4-Line 9. Then, we used emfjson-jackson plugin[16] to automatically transform SituationPrivacyWoTTD Ecore instance (Figure 5.7) into JSON document. This document is refined so, that, it refers to SituationPrivacy @context and finally represented as JSON-LD document. Finally, it is possible to make SituationPrivacy use existing vocabularies such as data privacy vocabulary,[17] context vocabulary,[18] or any other vocabulary deemed appropriate.[19]

Notes

1 www.weforum.org/projects/accelerating-the-impact-of-iot-technologies.
2 tinyurl.com/2p96m4fy.
3 www.businessinsider.com/iot-security-privacy.
4 www.opengroup.org/iot/odf/index.htm.
5 Readers can consult www.w3.org/TR/wot-thing-description for an extensive description of WoT TD.
6 www.weforum.org/agenda/2019/10/ageing-economics-population-health.
7 "*Falls are the number one cause of injury in seniors. One-third of seniors fall every year and 2.3 million of them end up in the emergency room because of a fall*" [25].
8 www.eclipse.org/modeling/emf.
9 github.com/SOM-Research/wot-toolkit.

10 www.thingweb.io.
11 https://nodered.org.
12 flows.nodered.org/node/node-red-contrib-web-of-things.
13 github.com/node-red/node-red-nodegen.
14 github.com/kit-sdq/Ecore2OWL
15 abenna.github.io/SituationPrivacy
16 github.com/emfjson/emfjson-jackson.
17 w3.org/ns/dpv.
18 github.com/ocabgit/Three-LevelContextOntology.
19 schema.org.

REFERENCES

1. A. Abdalla Hamad, Q. Z. Sheng, W. Emma Zhang, and S. Nepal. Realizing an Internet of Secure Things: A Survey on Issues and Enabling Technologies. *IEEE Communications Surveys & Tutorials*, 22(2), 1372–1391, 2020.
2. M. R. Abdmeziem, D. Tandjaoui, and I. Romdhani. Architecting the Internet of Things: State of the Art. In A. Koubaa and E. Shakshuki, editors, *Robots and Sensor Clouds*, Chapter 3, pp. 55–75. Springer International Publishing, New York.
3. N. Backx. Model-Driven Architecture for Web Services Applications. Technical report, University of Twente, The Netherlands, and Universidade Federal do Espírito Santo, Brazil, 2004.
4. P. M. Barnaghi and A. P. Sheth. On Searching the Internet of Things: Requirements and Challenges. *IEEE Intelligent Systems*, 31(6), 71–75, 2016.
5. K. Belhajjame, N. Faci, Z. Maamar, V. A. Burégio, E. Soares, and M. Barhamgi. On Privacy-Aware eScience Workflows. *Computing*, 102(5), 1171–1185, 2020.
6. I. Berrouyne, M. Qamar Adda, J. M. Mottu, J. C. Royer, and M. Tisi. A Model-Driven Approach to Unravel the Interoperability Problem of the Internet of Things. In *Proceedings of the 34th International Conference on Advanced Information Networking (AINA'2020)*, Caserta, Italy, 2020.
7. J. Bézivin, S. Hammoudi, D. Lopes, and D. Jouault. Applying MDA Approach for Web Service Platform. In *Proceedings of the 8th International Enterprise Distributed Object Computing (EDOC'2004)*, Monterey, CA, 2004. IEEE Computer Society.
8. M. Blackstock and R. Lea. IoT Mashups with the WoTKit. In *Proceedings of the 3rd IEEE International Conference on the Internet of Things (IOT'2012)*, pp. 159–166, Wuxi, Jiangsu Province, China, 2012.
9. B. Bordbar and A. Staikopoulos. On Behavioural Model Transformation in Web Services. In *Proceedings of the 5th International Workshop on Conceptual Modeling Approaches for e-Business (eCOMO'2004) Held in Conjunction with the 23rd International Conference on Conceptual Modeling (ER'2004)*, Shanghai, China, 2004.
10. A. Boscarino. Passenger Name Records in the Framework of EU Principles of Data Protection. Technical report, GRIN Institute of Technology, 2018.
11. A. Bröring, A. Ziller, V. Charpenay, S. Schmid, A. S. Thuluva, D. Anicic, A. Zappa, M. P. Linares, L. Mikkelsen, and C. Seidel. The BIG IoT API: Semantically Enabling IoT Interoperability. *IEEE Pervasive Computing*, 17(4), 41–51, 2018.
12. N. Bugshan, I. Khalil, N. Moustafa, and M. S. Rahman. Privacy-Preserving Microservices in Industrial Internet of Things Driven Smart Applications. *IEEE Internet of Things Journal*, 10(4), 2821–2831.
13. Electronic Privacy Information Center. EU-US Airline Passenger Data Disclosure, 2016.

14. P. Charith, B. Mahmoud, K. B. Arosha, A. A. Muhammad, A. P. Blaine, and N. Bashar. Designing Privacy-aware Internet of Things Applications. *Information Sciences*, 512, 238–257.
15. F. Ciccozzi and R. Spalazzese. MDE4IoT: Supporting the Internet of Things with Model-Driven Engineering. In *Proceedings of the 11th International Symposium on Intelligent and Distributed Computing (IDC'2017)*, Belgrade, Serbia, 2017.
16. A. Darko, T. Boris, and K. Tonimir. Interoperability and Lightweight Security for Simple IoT Devices. In *Proceedings of the 40th International Convention on Information and Communication Technology, Electronics and Microelectronics (MIPRO'2017)*, Opatija, Croatia, 2017.
17. S. K. Datta and C. Bonnet. Describing Things in the Internet of Things: From CoRE Link Format to Semantic Based Descriptions. In *Proceedings of the 2016 IEEE International Conference on Consumer Electronics-Taiwan (ICCE-TW'2016)*, Nantou County, Taiwan, 2016.
18. DZone. The Internet of Things, Application, Protocols, and Best Practices. Technical report, DZone, https://dzone.com/guides/iot-applications-protocols-and-best-practices, 2017 (visited in May 2017).
19. J. Eduardo Plazas, S. Bimonte, M. Schneider, C. de Vaulx, P. Battistoni, M. Sebillo, and J. C. Carlos Corrales. Sense, Transform & Send for the Internet of Things (STS4IoT): UML Profile for Data-centric IoT Aapplications. *Data & Knowledge Engineering*, 139, 101971, 2022.
20. M. F. Elrawy, A. I. Awad, and H. Hamed. Intrusion Detection Systems for IoT-based Smart Environments: A Survey. *Journal of Cloud Computing Advances Systems and Applications*, 7, 1–20, 2018.
21. G. Fortino, W. Russo, C. Savaglio, W. Shen, and M. Zhou. Agent-oriented Cooperative Smart Objects: From IoT System Design to Implementation. *IEEE Transactions on Systems, Man, and Cybernetics: Systems*, 11(48), 1939–1956, 2017.
22. M. A. Haque, S. Shetty, K. Gold, and B. Krishnappa. Realizing Cyber-Physical Systems Resilience Frameworks and Security Practices. In A. I. Awad, S. Furnell, M. Paprzycki, and S. K. Sharma, editors, *Security in Cyber-Physical Systems: Foundations and Applications*. Springer Nature, Berlin, pp. 1–38, 2021.
23. A. M. Hassan and A. I. Awad. Urban Transition in the Era of the Internet of Things: Social Implications and Privacy Challenges. *IEEE Access*, 6, 1–13, 2018.
24. J. H. Henrik Ziegeldorf, O. García Morchon, and K. Wehrle. Privacy in the Internet of Things: Threats and Challenges. *Security and Communication Networks*, 7(12), 2728–2742, 2014.
25. K. Hicks. The 5 Most Important Smart Tech Devices for Senior Safety. https://www.aplaceformom.com/caregiver-resources/articles/smart-tech-devices-for-senior-safety, December 2019.
26. I. Janggwan, H. K. Seong, and K. Daeyoung IoT Mashup as a Service: Cloud-Based Mashup Service for the Internet of Things. In *Proceedings of the 2013 IEEE International Conference on Services Computing (SCC'2013)*, Santa Clara, CA, 2013.
27. F. Jouault, F. Allilaire, J. Bézivin, and I. Kurtev. ATL: A Model Transformation Tool. *Science of Computer Programming*, 72(1–2), 31–39, 2008.
28. A. E. Khaled, S. Helal, W. Lindquist, and C. Lee. IoT-DDL–Device Description Language for the "T" in IoT. *IEEE Access*, 6, 1, 2018.
29. K. Le, S. K. Datta, C. Bonnet, and F. Hamon. WoT-AD: A Descriptive Language for Group of Things in Massive IoT. In *Proceedings of the 2019 IEEE 5th World Forum on Internet of Things (WF-IoT'2019)*, pp. 257–262, Limerick, Ireland, 2019.

30. T. Leppänen and J. Riekki. A Lightweight Agent based Architecture for the Internet of Things. In *Proceedings of the IEICE Workshop on Smart Sensing, Wireless Communications, and Human Probes*, Wuxi, China, March 2013.
31. Z. Maamar, M. Asim, K. Boukadi, T. Baker, S. Saeed, I. Guidara, F. Yahya, E. Ugljanin, and D. Benslimane. Towards a Quality-of-Thing Based Approach for Assigning Things to Federations. *Cluster Computing, Springer Nature*, 23(3), 1589–1602, 2020.
32. Z. Maamar, T. Baker, N. Faci, M. Al-Khafajiy, E. Ugljanin, Y. Atif, and M. Sellami. Weaving Cognition into the Internet-of-Things: Application to Water Leaks. *Cognitive Systems Research*, 56, 233–245, 2019.
33. Z. Maamar, M. Sellami, N. Faci, E. Ugljanin, and Q.Z. Sheng. Storytelling Integration of the Internet of Things into Business Processes. In *Proceedings of the Business Process Management Forum (BPM Forum'2018) Held in Conjunction with the 16th International Conference on Business Process Management (BPM'2018)*, Sydney, Australia, 2018.
34. A. Maaradji, H. Hacid, R. Skraba, A. Lateef, J. Daigremont, and N. Crespi. Social-Based Web Services Discovery and Composition for Step-by-Step Mashup Completion. In *Proceedings of the IEEE International Conference on Web Services (ICWS'2011)*, Washington, DC, 2011.
35. A. Martínez-Ballesté, P. A. Pérez-Martínez, and A. Solanas. The Pursuit of Citizens' Privacy: A Privacy-Aware Smart City is Possible. *IEEE Communications Magazine*, 51(6), 136–141, 2013.
36. J. Miller and J. Mukerji. MDA Guide Version 1.0.1. Technical report, Object Management Group, omg/2003-06-01, 2003.
37. A. M. Mzahm, M. S. Ahmad, and A. Y. C. Tang. Agents of Things (AoT): An Intelligent Operational Concept of the Internet of Things (IoT). In *Proceedings of the 13th International Conference on Intellient Systems Design and Applications (ISDA'2013)*, Bangi, Malaysia, 2013.
38. H. Orman. You Let That In? *IEEE Internet Computing*, 21(3), 99–102, 2017.
39. J. Phuttharak and S. W. Loke. A Review of Mobile Crowdsourcing Architectures and Challenges: Toward Crowd-Empowered Internet-of-Things. *IEEE Access*, 7, 1, 2019.
40. A. Qamar, A. Muhammad, Z. Maamar, T. Baker, and S. Saeed. A Quality-of-Things Model for Assessing the Internet-of-Thing's Non-Functional Properties. *Transactions on Emerging Telecommunications Technologies*, 33, 2019.
41. C. M. Sosa-Reyna, E. Tello-Leal, and D. Lara Alabazares. Methodology for the Model-driven Development of Service Oriented IoT Applications. *Journal of System Architecture*, 90, 15–22, 2018.
42. C. M. Sosa-Reyna, E. Tello-Leal, D. Lara-Alabazares, J. A. Mata-Torres, and E. Lopez-Garza. A Methodology-based on Model-Driven Engineering for IoT Application Development. In *Proceedings of the Twelfth International Conference on Digital Society and eGovernments (ICDS'2018)*, Rome, Italy, 2018.
43. X. Thang Nguyen, H. Tam Tran, H. Baraki, and K. Geihs. FRASAD: A Framework for Model-Driven IoT Application Development. In *Proceedings of the 2nd IEEE World Forum on Internet of Things (WF-IoT'2015)*, Milan, Italy, 2015.
44. S. Von Solms and S. Furnell. Human aspects of iot security and privacy. In A. I. Awad and J. Jemal Abawajy, editors, *Security and Privacy in the Internet of Things: Architectures, Techniques, and Applications*, pp. 31–55. John Wiley & Sons, Hoboken, NJ, 2022.
45. M. Weiser. The Computer for the 21st Century. *Newsletter ACM SIGMOBILE Mobile Computing and Communications Review*, 3(3), 3–11, 199.
46. A. F. Westin. Privacy and Freedom. *Washington and Lee, Law Review*, 25(1), 1–6, 1968.

47. R. Xiao, Z. Wu, and D. Wang. A Finite-State-Machine Model driven Service Composition Architecture for Internet of Things Rapid Prototyping. *Future Generation Computer Systems*, 99, 473–488, 2019.
48. J. Xu, K. Xue, S. Li, H. Tian, J. Hong, P. Hong, and N. Yu. Healthchain: A Blockchain-Based Privacy Preserving Scheme for Large-Scale Health Data. *IEEE Internet of Things Journal*, 6(5), 1–13, 2019.
49. F. Zambonelli. Key Abstractions for IoT-Oriented Software Engineering. *IEEE Software*, 34(1), 38–45, 2017.
50. Z. Zhang, M. C. Y. Cho, C. Wang, C. Hsu, C. Chen, and S. Shieh. IoT Security: Ongoing Challenges and Research Opportunities. In *Proceedings of IEEE 7th International Conference on Service-Oriented Computing and Applications (SOCA'2014)*, Matsue, Japan, 2014.

6 Protect the Gate
A Literature Review of the Security and Privacy Concerns and Mitigation Strategies Related to IoT Smart Locks

Hussein Hazazi and Mohamed Shehab
University of North Carloina at Charlotte

6.1 INTRODUCTION

With the advent of the Internet of Things, the word "smart" has become increasingly associated with the concept of taking a simple device that typically has basic features and enhancing them by adding multiple functions, as well as the ability to communicate with other devices. Although communicating with other devices makes those devices smart, they can also be more vulnerable to security and privacy issues because of their ability to communicate. An unauthorized manipulation of software or hardware in these devices can lead to the leakage of sensitive user information, as discussed in Kumar and Patel [22]. As Lin et al. explained in Ref. [25], smart devices with internet connectivity are considerably more vulnerable to remote attacks because attackers can download malware to them or directly access their networked control interfaces when connected to the internet. In the last few years, smart locks have emerged as a replacement for traditional locks that offer more features and enhancements. Approximately 0.42 billion U.S. dollars were spent on smart locks in 2016, according to a Statista report; however, by 2027, the market is expected to surpass 4 billion dollars [34]. As consumers gradually replace their traditional locks with smart locks, it is becoming increasingly important to investigate the security and privacy issues associated with smart locks. In fact, smart locks are estimated to have an even larger global market size if not for security and privacy concerns. According to their paper "Smart Locks for Smart Customers?", Hylta and Söderberg described a study conducted in London in 2017 on customers' adoption of smart locks. Sixty-four percent of the 54 respondents said they would hesitate to purchase a smart lock, with 50% citing security concerns as the reason [3].

 DOI: 10.1201/9781003199410-6

This chapter discusses the privacy and security issues related to smart locks, as well as mitigation strategies and platforms from both a researcher's and an end user's perspective. In the literature, there are several references analyzing commercially available smart locks, testing them for vulnerabilities in security and privacy, and proposing possible mitigations that we were able to explore and will be referencing throughout this chapter. In contrast, we found few research papers that examined the security and privacy of smart locks from the user's perspective. As a result, we believe there is a gap in the research on smart locks since there must be a user's input on the privacy and security of the device's main purpose which is to ensure the security and privacy of the household. To overcome this issue and get an understanding of what could be the end user's perspective regarding the privacy and security of smart locks, we reviewed research papers that discuss the privacy and security concerns and mitigation strategies related to **smart homes** from the perspective of the **end user.** Our rationale behind this is the fact that a smart lock is considered to be a smart home device and shares most of its characteristics including being able to connect to other smart home devices for automation purposes, keeping access logs, collecting private data, granting/revoking others access to the device, etc. Because smart locks share most of their characteristics with other smart home devices, they also share many security and privacy concerns and possible mitigation strategies. In this chapter, the following main objectives will be discussed:

- Identify the main smart locks security and privacy issues that were discussed in the literature, as well as explore some of the systems and platforms that were proposed to address them.
- Analyze the studies that examined the security and privacy issues of smart home devices and the mitigation strategies from the perspective of end users to identify the issues and mitigation strategies applicable to smart locks.
- Identify the research gaps in the field of smart locks privacy and security.

6.1.1 BACKGROUND

The Internet of Things has seen a tremendous amount of sophistication and diversity over the past two decades, ranging from applications for improving and automating services such as healthcare and smart manufacturing to smart home applications that give users more control over their home devices in order to improve human quality of life. A smart lock is a smart home device that was introduced as a replacement for the traditional lock [27]. A replacement that offers far more features beyond just locking and unlocking the door. Over the past few years, the smart lock market has grown and become more competitive, resulting in different designs and operational characteristics being introduced to the market [3].

6.1.2 ARCHITECTURE

Smart locks consist of three main components that are an electronically augmented deadbolt installed on the door, a companion mobile application installed on the user's smartphone, and a remote web server [36]. In order to control the smart lock, users must have an active account on the companion mobile application. As far as architecture is concerned, smart locks can use one of two types of network designs. Those two network designs are the Device-Gateway-Cloud (DGC) and Direct Internet Connection [17].

6.1.2.1 Device-Gateway-Cloud (DGC)

Since most smart locks store their data in the cloud, they require internet connectivity to connect to their remote servers in order to receive updates on access control instructions. As shown in Figure 6.1, A smart lock utilizing this architecture is not directly connected to the Internet. It is, however, possible for such locks to retrieve the necessary information from the cloud in two different ways. One approach involves connecting the lock to the user's smartphone via a local wireless channel, such as Bluetooth low power (BLE) [36], and then using the smartphone's internet connection as a gateway for connecting to the lock's remote servers and retrieving the necessary information and updates. The downside of this approach is that to control and use the features of the smart lock, the user must be within Bluetooth range of the lock. To overcome this issue, smart lock manufacturers that construct their locks using the DGC architecture typically provide the users with the option to purchase a Wi-Fi bridge (usually sold separately) as an alternative method of connecting the smart lock to the server without relying on the smartphone's internet connection. The Wi-Fi bridge communicates with the smart lock via Bluetooth or another short-range wireless technology, such as Z-Wave. Using this setup, smart locks can connect to the cloud as long as the Wi-Fi bridge is working correctly, allowing users to control their locks remotely using their smartphones without being within Bluetooth range.

6.1.2.2 Direct Internet Connection

Smart locks that use this architecture are equipped with a Wi-Fi modem embedded in the lock which allows them to connect to the home's Wi-Fi network [36]. The smart lock is therefore capable of connecting directly through Wi-Fi to the remote servers in order to retrieve the necessary information and updates, as illustrated in Figure 6.2. Using the companion application on their smartphone, users can remotely control the lock. However, since smart locks that use the direct internet connection

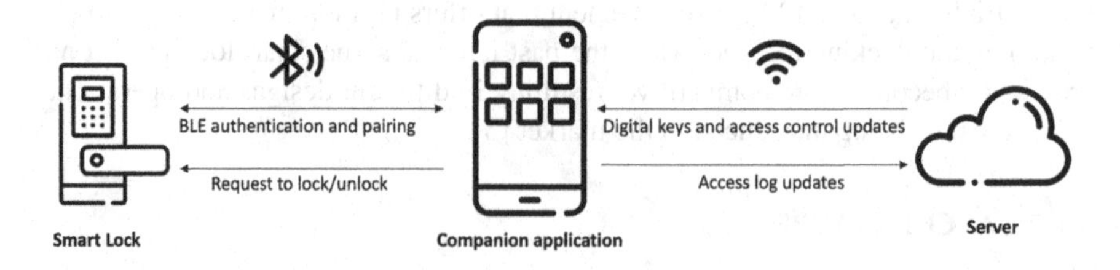

Figure 6.1 Device-gateway-cloud architecture.

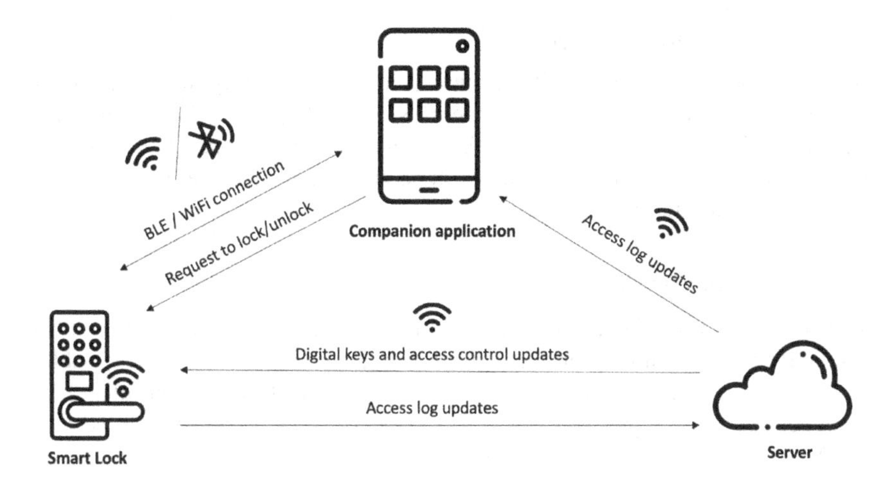

Figure 6.2 Direct internet connection architecture.

architecture can connect directly to the internet, all communication with the cloud is done through the lock's Wi-Fi connection only, which poses the possibility that the user will be locked out if the user does not have access to the internet.

6.1.3 CAPABILITIES

6.1.3.1 Locking/Unlocking the Door

Locking and unlocking the door is the main capability of every smart lock. Most smart locks require the user to lock or unlock the door by pressing a button in their companion application, by entering a PIN, or by using biometric characteristics such as face recognition or fingerprints. When the user's smartphone is within Bluetooth range of the door, some smart locks automatically unlock the door. If the door is not manually locked by the user, the door will automatically lock after a predetermined number of seconds unless the user manually locks it. Other smart locks, such as the Kevo Kwikset, require the user to touch the lock in order to unlock it, which is only possible if the smartphone is within Bluetooth range of the lock [28].

6.1.3.2 Exchanging Electronic Keys

One of the most significant distinctions between smart locks and traditional locks is that smart locks' owners are able to grant other users access to the lock as well as revoke that access electronically. Those electronic keys are tokens that contain the necessary information about the owner who is granting or revoking access, the user who is receiving the key or whose key is being revoked, as well as environmental information such as the duration of the access [28].

6.1.3.3 Keeping Access Logs

Every interaction with the smart lock, such as locking/unlocking the door, sending a digital key to someone, revoking someone's digital key, or updating someone's access level, is kept in an access log (also known as activity feed). The access log along with the timestamp of every entry can be viewed by the owner and those who have admin capabilities [28].

6.1.3.4 Vacation Mode

Some smart locks offer a vacation mode as a way of increasing the security of the household while the family is on vacation for a certain period of time. When enabled, the vacation mode disables all users' access codes so that no one can unlock the door unless they use a physical key or use a specific code to disable the vacation mode [30].

6.1.4 ACCESS CONTROL

Most commercial smart locks use role-based access control mechanisms that have predefined access levels and allow the owner to set the dates and duration in which the users can operate the lock [36]. They have four main access levels, namely, owner, resident, recurring guest, and temporary guest [17]. A user with the owner access level can grant access to other users through electronic keys, revoke other users' access, read the access log, and lock/unlock the door at any time. A user with a resident access level, however, cannot provide access to other users, revoke their access, or view access logs, although they can operate the lock at any time. Users with recurring guest access can, on the contrary, only operate the lock during fixed times set by the lock owner (e.g., every Thursday from 8 am to 10 am). Lastly, a user with a temporary guest access level may operate the lock for a period of time that is predetermined (e.g. 24 hours).

6.1.5 AUTHENTICATION AND AUTHORIZATION

Smart lock users must create an account on the lock's website or companion app in order to be able to operate the lock using their smartphones. Their login credentials will serve as an authentication mechanism to ensure that only users with the correct credentials can operate the lock. Users can access their accounts from any device that has the lock's companion app installed [35]. The level of authorization depends on the access level associated with the user's digital key. The access list of the smart lock is typically stored in the cloud; therefore, each time the user attempts to operate the lock, the access list must be retrieved. Locks with a Wi-Fi modem can directly contact the cloud to determine whether a particular user is authorized to operate the lock at a particular time and date based on their access level. Conversely, smart locks that follow a DGC architecture rely on the smartphone or Wi-Fi bridge as a gateway to retrieve authorization information from the cloud to determine whether the user should be permitted to operate the lock at a specific time and date [17].

6.2 THE PRIVACY AND SECURITY OF SMART LOCKS

The advancement of technology and the increasing adoption of smart home devices and smart locks have resulted in the publication of numerous research papers that discuss the privacy and security issues associated with these technologies. They also presented some solutions and mitigation strategies that may prove beneficial to users of those devices in dealing with these concerns or manufacturers in improving the privacy and security aspects of those devices in the future. In this section, we will review research papers that focus on the security and privacy issues that are mainly related to smart locks. Our discussion will also include other research papers that address security and privacy issues related to smart homes, since smart locks are an integral part of smart homes and are capable of communicating with other smart devices, which allows them to inherit some of the security and privacy issues associated with smart homes.

6.2.1 SMART LOCKS PRIVACY AND SECURITY FROM THE PERSPECTIVE OF RESEARCHERS

In many cases, the concepts of privacy and security overlap, as a security threat can also be considered as a privacy threat, and this overlap is even more pronounced in the area of the Internet of Things [32]. Therefore, an attack on smart locks that allows an adversary to gain unauthorized access to a home poses a direct threat to a household's privacy.

6.2.1.1 Security and Privacy Concerns

6.2.1.1.1 State Consistency Attacks

Smart locks following a Device-Gateway-Cloud architecture are susceptible to state consistency attacks due to the lack of direct internet connection and the fact that they rely on the internet connection of the user's smartphone or a Wi-Fi bridge [14,17,35,36,38]. As these smart locks keep their access control lists in the cloud (the remote server), they use the internet connection of the smartphone or a Wi-Fi bridge to update the companion application installed on the user's smartphone with the latest updated access control instructions and lock state updates. As a result, this allows for the following two scenarios:

Revocation evasion: User X's access to the smart lock was revoked; however, user X has his/her smartphone disconnected from the internet and within Bluetooth connectivity range to the smart lock. User X can still operate the lock because the lock companion application that is installed on the smartphone does not have connection to the internet, which means that it does not get the most recent update of the access control list. Additionally, no evidence of user X operating the lock will appear on the access log on the smart lock owner's companion app since user X's smartphone is not connected to the internet, which means the lock's state (locked or unlocked) will not be updated on the cloud.

Access log evasion: User X still has access legitimate to the smart lock; however, he/she turns off their smartphone's internet connection and gets within Bluetooth connectivity range from the smart lock in order to operate the smart lock without allowing the owner of the smart lock and those who have admin capabilities to know that user X has operated the smart lock. Again, since user X's smartphone is not connected to the internet, the smart lock's companion application that they have installed on their smartphone will not be able to send state updates to the cloud so that they can be added to the access logs.

To avoid those issues, smart locks can be easily instructed to not respond to the user's instructions while the user's smartphone is offline and ask the user to reconnect to the internet in order to be able to operate the lock; however, although this guarantees more security, it negatively affects the availability of the smart lock which might cause user frustration about being locked out of his/her home in case of a network outage or a lock-server connection issue [9].

6.2.1.1.2 *Relay Attacks*

A number of studies [17,28,29,35,36] have demonstrated that smart locks are susceptible to relay attacks, which are a form of Man-in-the-Middle (MitM) attack. As depicted in figure 6.3, a relay attack consists of two attackers (A1 and A2), one of whom is within proximity of the smart lock (A1), while the other is within proximity of an authorized user. The attacker who is close to the smart lock (A1) uses his Bluetooth relay device to capture the Bluetooth authentication message and relays it to the other attacker who is near the smart lock's legitimate user (A2). A2 broadcasts the signal and intercepts the response message from the user's phone and relays the message to A1, who in turn broadcasts the message to the smart lock to get it unlocked. Relay attacks can only be a threat to smart locks that follow a DCG architecture since these devices rely on Bluetooth to connect to the user's smartphone. Furthermore, because of how this attack works, the lock needs to also either have the auto-unlock feature ON or have a touch-to-unlock feature for this attack to be successful since the attackers will need to initiate the unlocking process for it to actually unlock [21].

6.2.1.1.3 *Unauthorized Unlocking*

It has been shown that smart locks can be susceptible to unauthorized unlocking depending on their design as well as their authentication and authorization

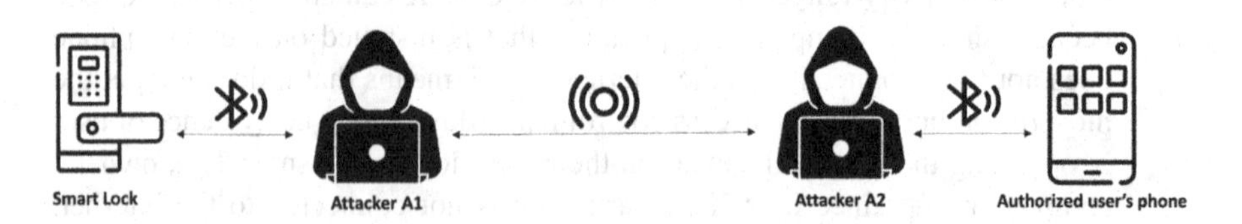

Figure 6.3 Relay attack.

mechanisms [13]. For example, smart locks that allow the user to unlock the door through a keypad are susceptible to shoulder surfing attacks. Furthermore, smart locks that store user data, as well as the lock's usage information, on a cloud server can also be compromised by hackers, especially if the authentication mechanism favors the convenience of the user over the safety of the system [20]. In addition, it is possible to gain unauthorized access to smart locks through lost or stolen smartphones, especially if the smartphone is not screen locked or the lock is set to automatically unlock when it is within Bluetooth connectivity range [33].

6.2.1.1.4 Tenant Privacy

In most rental properties, landlords are beginning to replace traditional locks with smart locks for a variety of reasons, including providing tenants with easier access and avoiding the necessity of changing the locks when the tenant moves out. However, the landlords or some of the employees at the property management company have full access to the lock and can view the tenant's usage information, access logs, and list of guests, friends, or family members who have access to the lock along with when each one of them enters the home or leaves it [8]. An invasion of privacy may result from this practice since it can result in issues such as profiling and surveillance, which may not even be apparent to the tenant. As a matter of fact, in 2019, tenants living in a New York City apartment complex sued their landlord because they were forced to switch from traditional locks to smart locks. Due to the amount of information that the landlord could obtain about them based on how they used the smart lock, they felt that the smart lock was a threat to their privacy. The tenants were successful in their lawsuit, and the landlord was forced to reinstall the traditional locks [10].

6.2.1.1.5 Companion Applications Vulnerabilities

As a result of a security analysis performed in Ref. [38], it has been concluded that the August smart lock is vulnerable to handshake key leakage attacks since the companion app stores the handshake key in plaintext and unencrypted on the smartphone where the companion app is installed. The attacker is therefore able to extract that handshake key with a rooted or jailbroken phone. By exploiting this vulnerability, an attacker is able to operate an August smart lock simply by executing the control program posted in the Augustctl Github repository. Knight et al. [21] discussed another vulnerability relating to companion apps. According to the analysis of the security of the Master Lock Bluetooth padlock, a vulnerability was found with the temporary codes that were provided to users. The companion app's interface advertises that the temporary code is valid for only four hours, whereas the message generated to share the code indicates that it will expire after eight hours.

6.2.1.2 Proposed Solutions

In Ref. [17], Ho et al. proposed an eventual consistency model as a means to mitigate the danger of state consistency attacks in smart locks that operate on a DGC architecture. With an eventual consistency model, the user's companion application retrieves a signed and updated access list from the server and sends it to the lock upon each attempt to operate the lock. To avoid replay of old access lists, the list should include a timestamp or an incrementing version number. Every time someone with an internet connection attempts to operate the lock, the lock will receive an updated access list. It is, therefore, possible for an attacker to perform a revocation evasion state consistency attack only until a legitimate user operates the lock, which then sends the lock an updated access list indicating that the attacker's access has been revoked. Furthermore, access log evasion state consistency attacks can be mitigated through the use of eventual consistency since the lock will maintain a copy of the latest log entries and will push them to the server whenever the lock is operated by a legitimate user who has a connection to the internet, making the logs accessible to the lock owner.

Xin et al. proposed an attribute-based access control (ABAC) framework for smart locks utilizing the DGC architecture in Ref. [36]. The proposed system is designed to eliminate state consistency attacks, cascading deletion of permissions, and unauthorized unlocking, as well as allow smart lock administrators to create and manage more fine-grained access control policies. While most of the smart locks sold commercially follow a role-based access control (RBAC) model, ABAC follows an attribute-based approach that comprises four attributes: subject (S), object (O), permission (P), and environment (E) where subjects represent the lock's users. Another important distinction between the ABAC system and most of the commercially available smart locks is that the administrator of a smart lock that uses the ABAC system needs to be within connection range to the lock in order to give other users access permissions or revoke their access privileges because the policy set is kept inside the lock as opposed to keeping it on the cloud like most commercially available smart locks do which makes them susceptible to state consistency attacks. As much as this feature increases the security of the ABAC systems, users might find it less convenient to have to be close to the smart lock in order to manage access control policies as opposed to being able to manage them remotely. Moreover, the proposed system attempts to solve the cascading deletion of permissions by adding the sRole attribute to the access control policy, which assigns the same sRole value to the person who created the permission and the person who received it. Due to their shared sRole value, the owner of the lock is able to revoke both privileges using the sRole.

Silva et al. present a method for protecting tenants' privacy and security when using online platforms for hospitality services such as Airbnb [8]. In the proposed system, guests are able to have full control over the smart lock during their stay, so that no one else can revoke their access or check the status of the lock during the term of their contract. In this system, access control rules are managed by smart contracts, which are provided by the Ethereum blockchain platform. Once the guest's contract expires, the system automatically revokes his access to the lock so that the host can regain control.

The authors of Ref. [2] discuss the smart lock's vulnerability to Man-in-the-Middle attacks and propose a method for improving the security of smart locks against such attacks. This proposed system utilizes both cryptography and image steganography to conceal the data being transferred and deceive those who attempt to intercept the message. When the user uses the lock's companion app to enter the passkey, it gets encrypted using AES encryption and then the encrypted cipher text gets encoded into an image. The image is then sent over Bluetooth to the server on which the cipher text will be decoded from the received image. If the passkey is valid, the door unlocks; otherwise, it remains locked.

The authors of Ref. [29] proposed SecSmartLock (Secure Smart Lock), a framework that incorporates an architecture and a secure communication protocol. The purpose of this framework is to enhance the security of smart locks and mitigate issues related to revocation evasion and access log evasion that are common among smart locks that are based on DGC architecture. There are five main components of the proposed framework (SecSmartLock), namely, the owner, the smart lock, an authorized user, a camera, and a server. In this framework, log evasion is prevented because all interactions with the smart lock are immediately transmitted to the camera that is connected to the lock via Bluetooth. The camera sends this information along with a video recording to the server. This framework, however, will not be sufficient to solve the issue of revocation evasion without an addition to the mechanism. The server and the smart lock will share a secret random nonce called the central nonce. Every time the user is trying to unlock the smart lock, he has to receive the encrypted central nonce from the server and send it to the smart lock along with the decrypted authentication nonce. CCA-secure secret key encryption is used to encrypt the central nonce using the shared secret key. The lock then decrypts the central nonce and compares it to the locally stored central nonce as well as compares the decrypted authentication nonce to the one locally stored on the lock. Only if they match can the user unlock the smart lock. Consequently, the user is always required to have an internet connection in order to receive the encrypted central nonce from the server, which is required at the end of the authentication process. Unless the user's smartphone is connected to the internet, it will not receive the encrypted central nonce. The server will cease sending the encrypted central nonce to the user if the owner revokes the user's access. The design prioritizes security over availability, which may result in legitimate users who do not have access to the internet being locked out of the system. To overcome this, the owner of the lock is provided with a master PIN to manually send to the users. This will enable them to unlock the lock at any time without the need to go through the normal authentication process.

6.2.2 SMART HOMES PRIVACY AND SECURITY FROM THE PERSPECTIVE OF THE END USER

A smart lock is considered to be an integral part of most smart homes. Additionally, smart locks share many of the characteristics of smart home devices, including the ability to be controlled by a companion app, the ability to be connected to other smart home devices for automation purposes, the ability to keep access logs, the ability to collect private data, the capability of being part of the home network, the ability to control the device remotely, and the capability of granting remote access to other users. As a result, smart locks share many of the same security concerns and mitigation strategies that apply to other smart home devices. Therefore, we reviewed research papers that explored the security and privacy issues associated with smart home devices and how end users choose to mitigate them. The purpose of this section is to introduce the end users' concerns and mitigation strategies that were discussed in those papers and can be applied to smart locks.

6.2.2.1 Security and Privacy Concerns

6.2.2.1.1 Data at Risk in the Cloud

Regardless of which architecture a specific smart lock follows, whether it is a Device-Gateway-Cloud architecture or a direct internet connectivity architecture, the vast majority of these locks are designed to store usage logs, customer information, and access control instructions in the cloud rather than on the lock itself. Although this can be considered as a feature that allows the customer to remotely control the lock and receive real-time notifications, several studies [5,16,31,37,40,42] show that some users are concerned that their information may be accessed by hackers or unauthorized individuals. In Ref. [31], 17 of the 23 participants expressed concern about data breaches. Furthermore, 7 of the 25 participants expressed their preference in Ref. [37] that their smart home devices collect and process data locally rather than sending it to the cloud. In the same study, nine participants indicated that they wish to have explicit control over what information is collected by the cloud and also to be able to delete it. Based on the results of Ref. [42], 39 of the 42 participants believed that the data collected by their smart home devices were not secure, and 27 of the 42 participants were concerned that hackers could access that information. It should be noted, however, that the data collected by smart home devices differ depending on the type of device. There is a tendency for smart home owners to be more concerned about the data collected by audio/video devices than other devices [41], but some of them are also aware that other devices can also provide enough information that might compromise their security and privacy. A person who has access to the data collected by a smart lock is capable of operating the lock and invading the privacy of those who use it.

6.2.2.1.2 Data Collection and Mining

Smart locks collect data such as the users' full names, locations, when they are at home and when they are not on a daily basis, and information regarding other smart devices interacting with the smart lock, allowing parties with access to this data to create profiles for those users for targeted advertising, government surveillance, etc. The results of several studies [5,12,16,19,31,37,40–42] suggest that some users of smart home devices are concerned that the data collected about them could be misused. As a threat associated with smart home devices, 26% of the participants in Ref. [31] mentioned "improper use and sharing of their data". In Ref. [16], 40% of the participants believe that the data collected by the smart home devices is being sold to third parties, while 45% are concerned about household profiling. Different participants in Ref. [41] had different opinions regarding the collection of data based on who will be using it. For example, there was less concern about the manufacturer of the smart device accessing their data than they were about the internet service providers (ISPs) having access to their data. Nevertheless, they were most concerned about the government having access to the data. A mixed opinion was expressed by the participants in Ref. [41] regarding advertisers. About 55% of the participants were not concerned about advertisers accessing their data and receiving targeted advertisements, while the remaining participants were somewhat concerned. It is worth noting that over 70% of the participants in Ref. [16] believe that the manufacturers of smart home devices are partially responsible for any privacy or security issues related to smart home devices, as manufacturers possess considerable control over the protocols used to collect, transmit, and receive data in smart home environments.

6.2.2.1.3 Multi-User Challenges

Typically, smart home devices are controlled or accessed by multiple users with different types of relationships. As an example, the owner of a smart lock may grant access to that lock to family members living in the house or visiting the house. It is also possible to share a smart lock with roommates who are not from the same family. Each of these scenarios presents its own set of challenges and issues. A common example of such issues is access imbalance, in which those with a higher level of access are able to control how others in the household use smart devices. The purpose of the study conducted by Zeng and Roesner [39] was to discuss issues related to the use of smart devices within a household, as well as to determine what functionalities and preferences an end user who lives in a multi-user environment requires to ensure increased privacy and security. As an example, some participants required the capability of preventing guests from remotely controlling devices that they have access to as well as limiting their ability to use a specific device while physically present in order to feel more comfortable in a multi-user environment [39]. This could be applied to smart locks because it might not be desirable for a homeowner to allow his guests who have access to the smart lock to be able to operate the lock remotely. The use of smart home devices may also cause tension and direct conflict among household members over who has access to the house and who does not [11]. In Ref. [11], Geeng and Roesner discussed situations in which tension or conflict arises between household members due to their differing expectations regarding the use of smart devices. Within a household, such conflicts can arise in four types of relationships: the relationship between partners, the relationship between parents and children, the relationship between guests and homeowners, and the relationship between roommates. For example, there was some tension between one of the participants in the study and his girlfriend because they had different opinions on whether or not the cleaning lady should have access to the house, which is a smart lock functionality. In addition, the study showed that smart home devices can be used by parents as a tool to manage their children as well as set limits and specific schedules. A conflict can also arise between homeowners and their guests or non-occupants due to how some guests interact with the smart home devices in a manner they do not appreciate. A tension may also arise between roommates over how smart home devices should behave, such as who should control the thermostat [11]. Smart locks can contribute to tensions that may occur within a multi-user environment when one of the roommates does not agree with the fact that another roommate has provided a digital key or access code to another person whom the first roommate does not trust or get along well with. It was mentioned in Ref. [7] by some participants that communicating with other people who are also using smart home devices is an effective way of addressing their security and privacy concerns within a multi-user home environment. This will help them understand how they feel and what can be done to protect their personal information as well as the privacy of others.

6.2.2.1.4 *Network Attacks*

A major concern for owners of smart homes is the possibility of network attacks that may allow unauthorized access to smart home devices connected to the wireless network [16,31,37,40]. A total of 48% of the participants in Ref. [31] expressed concerns about the possibility of their Wi-Fi being hacked and remotely controlled, allowing adversaries to steal their personal information. If a hacker is able to gain access to their location information and smart lock access codes, then they will also be put in danger of their physical safety. Based on the results of Ref. [38], 3 of the 25 participants believe that an additional security feature for smart home devices should consist of "network intrusion detection."

6.2.2.1.5 *Insecure Devices*

In recent years, there has been a significant increase in the number of commercially available smart locks, as well as smart home devices in general, from different companies manufactured in different countries that are available on the market. A consumer may feel overwhelmed by the various options available when they are shopping for a new smart home device and may find it difficult to compare the various security features and privacy settings of various devices. Almost 27% of the participants in Ref. [40] are concerned that the smart home devices they purchase may not be sufficiently secure, and 20% are concerned that their smart home devices are malicious.

6.2.2.1.6 *Physical Safety*

There is a possibility that a compromised smart home device could pose a threat to the safety of family members within the household in terms of their physical safety. As a result, it is natural to assume that physical safety will be one of the most important concerns for users of smart homes. The significance of this is especially true when it comes to smart locks since a compromised smart lock may allow unauthorized entry into the home, which may result in household members being physically attacked by intruders if the lock is compromised. As per a study published in Ref. [16], over 41% of the participants expressed concern about their physical safety if their smart home devices were compromised or were able to collect sensitive information about them that could put them at risk.

6.2.2.2 Mitigation Strategies

Smart home devices, including smart locks, do not become obsolete simply because they have privacy and security issues. Ultimately, these devices provide a high level of convenience and reliability. As a result, some users choose to implement mitigation strategies rather than completely ignore smart home devices due to privacy and security concerns. Table 6.1 provides an overview of the various mitigation strategies employed by participants in four research papers [15,16,31,40] to address security and privacy issues associated with their smart locks.

6.2.2.2.1 Self-censoring

Self-censoring takes many different forms, such as deciding to not provide the smart device with additional information beyond what is necessary to complete the basic tasks, for example, by using nicknames when giving access to family members or guests. Thus, even if another party gained access to this information, it would not be able to identify that individual. Self-censorship can also be achieved by choosing to not utilize certain device functionalities, such as by turning off the "auto unlock" feature on smart locks to avoid the smart lock constantly checking your phone's location information to unlock your door before you approach; 50% of the participants in Ref. [31] and almost 20% of the participants in Ref. [16] stated that they use at least one form of self-censoring as a mitigation strategy to deal with privacy and security concerns related to their smart home devices.

6.2.2.2.2 Device Selection

Prior to purchasing a smart lock, conducting research to select one with a strong privacy and security feature as well as constant updates and maintenance can also serve as a mitigation strategy. Since different smart locks have different features and network infrastructure, they may be more vulnerable to specific security threats than others. For example, 17% of the respondents in Ref. [16] indicated that they take privacy and security into account when purchasing a smart home device.

Table 6.1
Comparing the Use of Mitigation Strategies Based on Four Research Papers [15,16,31,40]

Mitigation Strategy	Mitigation Technique	Number of Participants	Total
Non-technical mitigation	Self-censoring	25	37
	Device selection	12	
Technical mitigation	Network configuration	26	75
	Configuring device options	16	
	Authentication	33	

Combined, the four studies have a total of 93 participants.

6.2.2.2.3 *Network Configuration*

A number of studies [15,16,31,40] describe how participants configure their home networks to mitigate privacy and security concerns relating to their smart home devices. For example, using a separate network dedicated to smart home devices is one of the most effective methods of configuring home networks to increase their level of security and privacy. This way, smart home devices will not be affected by attacks on other electronic devices, such as smartphones or personal computers, and vice versa. However, it is important to note that only a small percentage of the participants in each of the three studies that discussed keeping smart home devices on a separate network had actually implemented this mitigation strategy. Additionally, the installation of virtual private networks (VPNs) and monitoring network traffic are two other forms of network configuration that smart home users implement to mitigate their privacy and security concerns.

6.2.2.2.4 *Configuring Device Options*

Most smart home devices are configured by default for ease of use and convenience, rather than for security, which is why some users opt to change their configurations based on their preferences. Twenty-nine percent of the participants in Ref. [16] reported configuring the settings on their smart home devices in order to increase the security and privacy of their devices. A smart lock, for example, may be configured so that it automatically unlocks when the user enters the neighborhood if the user prefers a more convenient experience. This may be more convenient, but is less secure since someone may be in the vicinity of the door and may gain access to the house before the owner does. Additionally, it allows the lock to continuously request the location of the homeowner's smartphone. Thus, some users choose to disable this feature in order to protect their privacy and security.

6.2.2.2.5 *Authentication*

As a means of increasing the security of smart home devices and preventing them from becoming easy targets for hackers, it is very effective to use strong authentication methods such as multi-factor authentication (MFA) or a strong password to protect user accounts associated with those devices [15,16,31]. As a matter of fact, 39% of the participants in Ref. [31] used it as a mitigation strategy to address their privacy and security concerns with respect to smart home devices. As mentioned above, not all smart home devices, including smart locks, support MFA, which highlights the importance of selecting the smart home device that offers the most security options.

6.3 RESEARCH GAPS

While many studies have examined the security and privacy issues associated with smart locks as well as proposed novel and viable mitigation strategies for resolving these challenges, we observed that the literature still lacks studies that examine the privacy and security issues associated with smart locks from the end users' perspective. It is rare in the literature to find work that discusses these issues from the perspective of the end users of smart locks. As a result, we believe that future research papers should address the following topics, such as exploring the privacy and security concerns associated with smart locks from the perspective of end users, examining and evaluating the mitigation strategies used by the end user to address these concerns, and assessing the level of familiarity of the end user with security and privacy issues associated with smart locks.

Other research papers have explored privacy and security issues related to smart home devices from the perspective of end users, as discussed in Section 9.2.2 of this chapter; however, it's also important to conduct similar studies that focus primarily on smart locks as the purpose of using smart locks and the types of data that smart locks collect may be different compared to other smart home devices which may result in different security and privacy concerns unique to smart locks. This is also true for mitigation strategies that can be used to deal with those unique privacy and security concerns related to smart locks. As an example, installing a video doorbell can serve as a mitigation strategy to address privacy and security concerns regarding smart locks, but this is not applicable to most other smart home devices. Therefore, and due to the fact that different smart home devices can have different end users' privacy and security concerns and mitigation strategies, many other studies were published that focus on specific smart home devices such as Refs. [1,4,6,18,23,24,26] that focus specifically on the privacy and security concerns related to smart speakers and virtual assistants from the perspective of their end users. Furthermore, the importance of discussing such topics from the viewpoint of the smart lock users can be explained by the fact that the privacy and security concerns of end users must be taken into account so that meaningful changes can be made in the future to the smart lock system and design to ensure that those improvements cater to the user's needs. Additionally, it is necessary to examine the mitigation strategies that smart lock users use to address their privacy and security concerns in order to systematically evaluate their effectiveness and weaknesses. Lastly, evaluating the end user's familiarity with security and privacy issues associated with smart locks is imperative, as smart locks are primarily responsible for the security and privacy of an entire house, as well as the security and privacy of its inhabitants and their belongings. In light of this, it is imperative that users of smart locks have a high level of awareness of the security and privacy challenges associated with smart locks so that they can plan to mitigate those challenges while also enjoying the convenience and ease of use associated with smart locks.

6.4 CONCLUSION

Increasing the security and privacy of smart locks has a direct impact on improving the security and privacy of the home and its inhabitants. In order to improve the security and privacy of smart locks, it is necessary to investigate privacy and security issues related to commercially available smart locks, as well as the mitigations proposed to mitigate these issues. In this chapter, we examined those issues and mitigation strategies from the perspective of both researchers and end users of smart home devices. Moreover, we identified the research gaps when it comes to studies on the privacy and security of smart locks. Future research should focus on exploring the security and privacy issues associated with smart locks from the viewpoint of end users, as well as evaluating the mitigation strategies end users employ to address and cope with their concerns.

REFERENCES

1. Abdi, N., Ramokapane, K. M., & Such, J. M. (2019). More than smart speakers: Security and privacy perceptions of smart home personal assistants. In *Fifteenth Symposium on Usable Privacy and Security (SOUPS 2019).*
2. Bapat, C., Baleri, G., Inamdar, S., & Nimkar, A. V. (2017, September). Smart-lock security re-engineered using cryptography and steganography. In *International Symposium on Security in Computing and Communication* (pp. 325–336). Springer, Singapore.
3. Hylta, S. B., & Söderberg, P. (2017). Smart locks for smart customers? A study of the diffusion of smart locks in an Urban Area [Master's thesis, KTH Royal Institute of Technology, School of Industrial Engineering and Management].
4. Chalhoub, G., & Flechais, I. (2020, July). "Alexa, are you spying on me?": Exploring the effect of user experience on the security and privacy of smart speaker users. In *International Conference on Human-Computer Interaction* (pp. 305–325). Springer, Cham.
5. Chhetri, C., & Motti, V. G. (2019, March). Eliciting privacy concerns for smart home devices from a user centered perspective. In *International Conference on Information* (pp. 91–101). Springer, Cham.
6. Cho, E., Sundar, S. S., Abdullah, S., & Motalebi, N. (2020, April). Will deleting history make Alexa more trustworthy? Effects of privacy and content customization on user experience of smart speakers. In *Proceedings of the 2020 CHI Conference on Human Factors in Computing Systems* (pp. 1–13).
7. Cobb, C., Bhagavatula, S., Garrett, K. A., Hoffman, A., Rao, V., & Bauer, L. (2021). "I would have to evaluate their objections": Privacy tensions between smart home device owners and incidental users. *Proceedings on Privacy Enhancing Technologies, 2021*(4), 54–75.
8. de Camargo Silva, L., Samaniego, M., & Deters, R. (2019, October). IoT and Blockchain for Smart Locks. In *2019 IEEE 10th Annual Information Technology, Electronics and Mobile Communication Conference (IEMCON)* (pp. 0262–0269). IEEE.
9. Doan, T. T., Safavi-Naini, R., Li, S., Avizheh, S., & Fong, P. W. (2018, August). Towards a resilient smart home. In *Proceedings of the 2018 Workshop on IoT Security and Privacy* (pp. 15–21).
10. Elizabeth Kim, Rosemary Misdary, Elizabeth Kim, J. G., George Joseph, & Elizabeth Kim. (2019, May 9). Hell's kitchen landlord sued for Keyless Entry System agrees to provide keys. Gothamist. Retrieved May 7, 2021, from https://gothamist.com/news/hells-kitchen-landlord-sued-for-keyless-entry-system-agrees-to-provide-keys.
11. Geeng, C., & Roesner, F. (2019, May). Who's in control? Interactions in multi-user smart homes. In *Proceedings of the 2019 CHI Conference on Human Factors in Computing Systems* (pp. 1–13).

12. Gerber, N., Reinheimer, B., & Volkamer, M. (2018, August). Home sweet home? Investigating users' awareness of smart home privacy threats. In *Proceedings of An Interactive Workshop on the Human Aspects of Smarthome Security and Privacy (WSSP).*
13. Gupta, S., Buriro, A., & Crispo, B. (2019, May). Smarthandle: A novel behavioral biometric-based authentication scheme for smart lock systems. In *Proceedings of the 2019 3rd International Conference on Biometric Engineering and Applications* (pp. 15–22).
14. Han, Z., Liu, L., & Liu, Z. (2019, May). An efficient access control scheme for smart lock based on asynchronous communication. In *Proceedings of the ACM Turing Celebration Conference-China* (pp. 1–5).
15. Haney, J. M., Furman, S. M., Theofanos, M. F., & Fahl, Y. A. (2019). Perceptions of smart home privacy and security responsibility, concerns, and mitigations. In *Poster at the 15th Symposium on Usable Privacy and Security (SOUPS 2019).*
16. Haney, J., Acar, Y., & Furman, S. (2021). "It's the company, the government, You and I": User perceptions of responsibility for smart home privacy and security. In *30th USENIX Security Symposium (USENIX Security 21).*
17. Ho, G., Leung, D., Mishra, P., Hosseini, A., Song, D., & Wagner, D. (2016, May). Smart locks: Lessons for securing commodity internet of things devices. In *Proceedings of the 11th ACM on Asia Conference on Computer and Communications Security* (pp. 461–472).
18. Huang, Y., Obada-Obieh, B., & Beznosov, K. (2020, April). Amazon vs. my brother: How users of shared smart speakers perceive and cope with privacy risks. In *Proceedings of the 2020 CHI Conference on Human Factors in Computing Systems* (pp. 1–13).
19. Kaaz, K. J., Hoffer, A., Saeidi, M., Sarma, A., & Bobba, R. B. (2017, October). Understanding user perceptions of privacy, and configuration challenges in home automation. In *2017 IEEE Symposium on Visual Languages and Human-Centric Computing (VL/HCC)* (pp. 297–301). IEEE.
20. Kassem, A., El Murr, S., Jamous, G., Saad, E., & Geagea, M. (2016, July). A smart lock system using Wi-Fi security. In *2016 3rd International Conference on Advances in Computational Tools for Engineering Applications (ACTEA)* (pp. 222–225). IEEE.
21. Knight, E., Lord, S., & Arief, B. (2019, August). Lock picking in the era of internet of things. In *2019 18th IEEE International Conference on Trust, Security and Privacy in Computing and Communications/13th IEEE International Conference on Big Data Science and Engineering (TrustCom/BigDataSE)* (pp. 835–842). IEEE.
22. Kumar, J. S., & Patel, D. R. (2014). A survey on internet of things: Security and privacy issues. *International Journal of Computer Applications, 90*(11), 20–26.
23. Lau, J., Zimmerman, B., & Schaub, F. (2018). Alexa, are you listening? Privacy perceptions, concerns and privacy-seeking behaviors with smart speakers. *Proceedings of the ACM on Human-Computer Interaction, 2*(CSCW) (pp. 1–31).
24. Lau, J., Zimmerman, B., & Schaub, F. (2018). Alexa, stop recording: Mismatches between smart speaker privacy controls and user needs. In *Poster at the 14th Symposium on Usable Privacy and Security (SOUPS 2018).*
25. Lin, H., & Bergmann, N. W. (2016). IoT privacy and security challenges for smart home environments. *Information, 7*(3), 44.
26. Malkin, N., Deatrick, J., Tong, A., Wijesekera, P., Egelman, S., & Wagner, D. (2019). Privacy attitudes of smart speaker users. *Proceedings on Privacy Enhancing Technologies, 2019*(4), 250–271.
27. Mamonov, S., & Benbunan-Fich, R. (2020). Unlocking the smart home: Exploring key factors affecting the smart lock adoption intention. *Information Technology & People*, 34(2), 835–861.

28. Palle, S. (2017). Smart locks: *Exploring security breaches and access extensions* [Doctoral dissertation, Oklahoma State University].
29. Patil, B., Vyas, P., & Shyamasundar, R. K. (2018, December). SecSmartLock: An architecture and protocol for designing secure smart locks. In *International Conference on Information Systems Security* (pp. 24–43). Springer, Cham.
30. Schlage Keypad Home User Manual, http://www.getsymon.com/myJSSImages/file/Manuals/SchlageProgramming.pdf.
31. Tabassum, M., Kosinski, T., & Lipford, H. R. (2019). "I don't own the data": End user perceptions of smart home device data practices and risks. In *Fifteenth Symposium on Usable Privacy and Security (SOUPS 2019).*
32. Tank, B., Upadhyay, H., & Patel, H. (2016, March). A survey on IoT privacy issues and mitigation techniques. In *Proceedings of the Second International Conference on Information and Communication Technology for Competitive Strategies* (pp. 1–4).
33. Tilala, P., Roy, A. K., & Das, M. L. (2017, November). Home access control through a smart digital locking-unlocking system. In *TENCON 2017-2017 IEEE Region 10 Conference* (pp. 1409–1414). IEEE.
34. Vailshery, L. S. (2021). "Global Smart Lock Market Size 2016–2027." *Statista*, 9 April 2021, www.statista.com/statistics/1117440/forecast-comparison-of-global-smart-lock-market-size/#:~:text=The%20market%20size%20of%20smart,according%20to%20different%20research%20agencies.
35. Viderberg, A. (2019). Security evaluation of smart door locks [Master's thesis, KTH Royal Institute of Technology, School of Electrical Engineering and Computer Science].
36. Xin, Z., Liu, L., & Hancke, G. (2020). AACS: Attribute-based access control mechanism for smart locks. *Symmetry*, *12*(6), 1050.
37. Yao, Y., Basdeo, J. R., Kaushik, S., & Wang, Y. (2019, May). Defending my castle: A co-design study of privacy mechanisms for smart homes. In *Proceedings of the 2019 CHI Conference on Human Factors in Computing Systems* (pp. 1–12).
38. Ye, M., Jiang, N., Yang, H., & Yan, Q. (2017, May). Security analysis of Internet-of-Things: A case study of august smart lock. In *2017 IEEE Conference on Computer Communications Workshops (INFOCOM WKSHPS)* (pp. 499–504). IEEE.
39. Zeng, E., & Roesner, F. (2019). Understanding and improving security and privacy in multi-user smart homes: A design exploration and in-home user study. In *28th USENIX Security Symposium (USENIX Security 19)* (pp. 159–176).
40. Zeng, E., Mare, S., & Roesner, F. (2017). End user security and privacy concerns with smart homes. In *Thirteenth Symposium on Usable Privacy and Security (SOUPS 2017)* (pp. 65–80).
41. Zheng, S., Apthorpe, N., Chetty, M., & Feamster, N. (2018). User perceptions of smart home IoT privacy. *Proceedings of the ACM on Human-Computer Interaction*, *2*(CSCW), (pp. 1–20).
42. Zimmermann, V., Bennighof, M., Edel, M., Hofmann, O., Jung, J., & von Wick, M. (2018). 'Home, smart home': Exploring end users' mental models of smart homes. *Mensch und Computer 2018-Workshopband.*

7 A Game-Theoretic Approach to Information Availability in IoT Networks

Abdallah Farraj
Texas A&M University - Texarkana

Eman Hammad
Texas A&M University

7.1 INTRODUCTION

The Internet of Things (IoT) is defined as the network of physical devices or "things" that are embedded with electronics, software, different kinds of sensors and actuators and are connected to the internet via heterogeneous access networks to enable "things" to exchange data with the manufacturer, operator, and/or other connected devices [24,31]. Industrial IoT (IIoT) is a specialized IoT device that is designed as part of industrial processes or products. Considering communication requirements, IIoT can be classified into three categories: sensors that mainly transmit measurements, actuators that mainly receive control commands, and sensors/actuators that combine the capabilities to transmit and receive. IIoT industrial use cases are vast, where they can perform sensing and actuation tasks with minimal human intervention [17,24].

Enabled by innovative technologies such as 5G/6G wireless connectivity, artificial intelligence, and machine learning, IoT will continue to find enormous opportunities in applications across a wide range of industry verticals. IoT is being widely deployed in industries such as healthcare, energy, transportation, and manufacturing, to name a few [16,25,26,33]. IoT use-cases are motivating a massive IoT adoption trend that predicts the connectivity of 75.44 billion devices by 2025 [33]. The increased utilization of IoT in critical and sensitive processes underscores the need to establish strong controls to ensure trusted and reliable operation. From a communication perspective, large-scale deployments of IoT can be supported by massive machine type communication (MTC) and machine-to-machine links; however, security might not be trivial at such scales.

DOI: 10.1201/9781003199410-7

When discussing IoT ecosystems, it is informative to mention the role of wireless sensor networks (WSN). WSN describe collections of networked wireless sensors that measure physical conditions and transmit readings to a central location. WSN be considered as a subset of IoT networks that enable improved industrial automation applications by providing means to monitor industrial environment conditions utilizing wireless IoT sensors (e.g., pressure and temperature sensors). WSN transfer collected data to a data hub for processing, observation, analysis, decision-making, and real-time close-loop control. Industrial WSN provide industrial plant operators with many benefits including quality and timely information required for decision-making, advanced distributed control, improved productivity, and better asset and process visibility [16,25].

In IoT networks, it is critical to consider the security objectives of confidentiality, integrity, and availability in the design and operation of such systems. Recalling that security objective of data availability ensures that authorized users can access information whenever required. In IoT networks, the reliability of the communication system directly affects data availability. Factors such as hardware failures, software downtime, human error, cyberattacks, and channel access opportunities can negatively impact data availability. To mitigate information availability concerns, system operators implement information security policies and security controls, including redundancies and backups, to ensure uninterrupted system operation and information availability. However, such approaches may not mitigate threats such as jamming and intentional electromagnetic interference attacks.

In this chapter, we consider the channel access problem for IoT devices in an industrial WSN [1,18,20,27,28] as we take another approach to address information availability in IoT networks. We utilize a cognitive communication system setup in which IoT devices are treated as the secondary users sharing the channel with the primary user. We focus our treatment on the case of IoT devices wanting to transmit sensor measurements to a common receiver unit without violating the primary user's outage probability requirements. A base station in an industrial plant can be modeled using the aforementioned common receiver unit. In contrast to recent work [26], we focus on uncoordinated channel access for IoT devices with the goal to satisfy information availability requirements.

Our model allows IoT devices (treated here as secondary users in a cognitive communication system) to transmit over a shared channel without coordinating their activity with other devices in the system [16,25]. The proposed model focuses on the transmission strategy of one secondary user regardless of the channel-access activity of other competing secondary users. This model is motivated by relevant industrial WSN settings where several different IoT devices require channel transmission. In such environments, the proposed approach would help alleviate the overhead of coordinating channel access between the users of the channel while not negatively impacting the performance of primary user.

7.2 RELATED WORK

Cybersecurity of IoT systems has been an active field of research and development for years. Recent works include [3–5,21,22,30,32]. Intrusion and anomaly detection schemes for IoT systems are proposed in Refs. [3,30], respectively. Vulnerability assessment for IoT-based smart homes is discussed in Ref. [4], and a framework for vetting IoT is described in Ref. [22]. A review of authentication and identity management practices for internet of healthcare things is provided in Ref. [21], and intelligent authentication of 5G healthcare devices is surveyed in Ref. [32]. Architectures and techniques for security and privacy of IoT systems are discussed in Ref. [5].

Spectrum-sharing cognitive communications is a promising technology to support mMTC as it improves spectrum utilization efficiency [23]. In a spectrum-sharing communication environment, unlicensed (also known as secondary) users adapt their communication parameters to be able to transmit over a wireless channel without violating a performance metric of the licensed (also known as primary) user of that channel. As such, cognitive communication systems can provide attractive solutions for IoT connectivity and mMTC while accommodating resource allocations [6,14,26,34]. Cognitive communication schemes can be extended to support the analysis of IoT transmissions in various communication environment including industrial plants.

A game-theoretic model is developed in this chapter to capture the secondary users' uncoordinated transmissions and investigate the impact of their transmission strategies on selected performance metrics. Specifically, the dynamical interactions between the secondary users in the industrial WSN are modeled using a 2×2 iterated game. Further, in a specific transmission interval, an IoT device of interest, denoted as the controller user, reacts to the transmission activity of opponent IoT devices from the previous transmission round. The select IoT device (i.e., controller user) adopts the game-theoretic strategy to transmit over the shared channel without coordinating its activity with other IoT devices in the system while complying with the quality of service (QoS) requirement, chosen here to be the primary outage probability.

The adopted iterated game models lend themselves to Markovian strategies, often called *zero-determinant* strategies [2,29]. Here, players with longer memory of the game history do not have a long-term performance advantage over other game players with shorter memories. With zero-determinant strategies, a game player can control its long-term payoff through exploiting the structure of the payoff matrix of the game. Thus, the proposed *zero-determinant* transmission strategy enables decentralized and uncoordinated channel access for IoT devices, which leads to a simplified scheduling algorithm with guarantees of meeting QoS requirements. Simplified scheduling algorithms, such as the one proposed in this chapter, better fit limited-resources IoT devices and would help extend battery life in battery-operated IoT.

The works in Refs. [8,9] demonstrate the feasibility of using physical-layer security practices in achieving security measures for IoT undergoing interference and eavesdropping attacks. Building on the promise of these recent approaches to security, this book chapter applies a physical-layer security approach to information availability through the use of uncoordinated IoT channel access.

To the best of the authors' knowledge, uncoordinated channel access for IoT device in a spectrum-sharing setting using zero-determinant-based transmission strategies has not been addressed before.

7.3 SYSTEM MODEL

WSN provide communication infrastructure to observe environment parameters and enable monitoring, analysis, and control. Recall that a sensing IoT device is typically resource constrained in communication, storage, processing power, and/or energy capabilities. Industrial networks could potentially include hundreds of sensing IoT devices; thus, given the resource-constrained nature of such devices, scheduling efficient channel access for IoT information availability is a problem of interest.

7.3.1 SPECTRUM-SHARING COGNITIVE SYSTEMS

In spectrum-sharing systems, a secondary user can simultaneously transmit over the channel along with the primary user under the constraint that secondary transmission activity does not deteriorate a QoS measure. QoS requirements often include having limits on the average or maximum secondary interference, limiting the outage probability of the primary user to some threshold, or having minimum signal-to-interference plus noise ratio (SINR) for the primary user's signal. The primary user's outage probability is chosen as the QoS constraint for spectrum-sharing systems in Refs. [7,10–13].

For multi-user spectrum-sharing systems, multiple secondary users want to transmit over the same channel in order to achieve data availability requirements. In this case, some scheduling authority could decide which user is scheduled to transmit over the shared channel; furthermore, the scheduling criterion might take metrics like fairness between users or channel conditions into considerations. For example, the secondary user with the weakest channel is assigned the channel in a multi-user environment in Ref. [19]; different scheduling schemes are investigated for spectrum-sharing systems Refs. [7,12].

In this chapter, we utilize a game-theoretic formulation to devise an uncoordinated strategy for IoT channel transmission for information availability while meeting the outage probability constraints of the primary user of the system.

7.3.2 PROBLEM STATEMENT

Consider a generic industrial WSN as illustrated in Figure 7.1. Here, the considered communication system includes a primary user (denoted as PU) that utilizes the channel to communicate data to a common receiver unit (denoted as RU). In this setting, PU could be the licensed user for the channel or a critical industrial device with stringiest QoS constraints. Also, consider multiple secondary users, collectively denoted as SUs, that want to communicate their own data to RU. SUs represent in this setup the multiple resource-constrained IoT sensors in the industrial network. Both PU and SU can transmit concurrently over the shared channel; however, concurrent

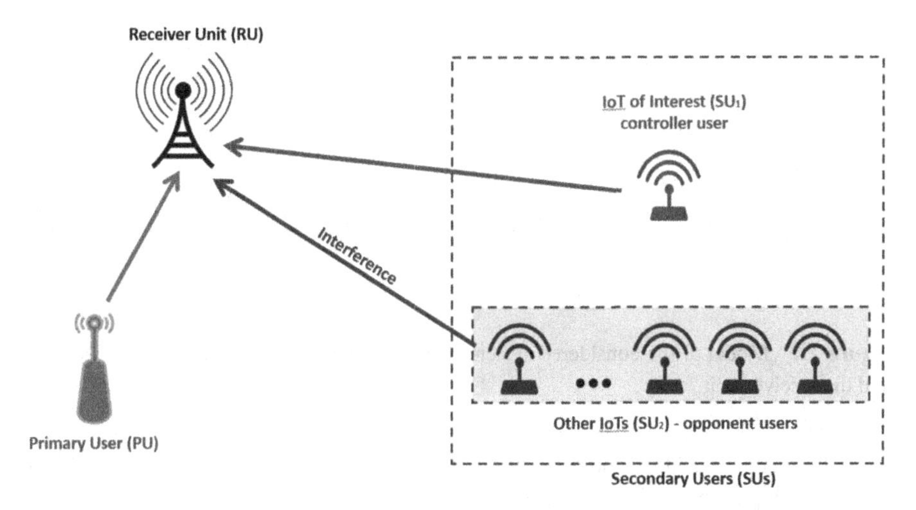

Figure 7.1 Problem setup for a generic industrial wireless sensor network showing IoT devices collaborating to access the shared channel.

secondary transmissions, when they happen, cause interference on PU's signal. The interference could be sensed at RU which affects the QoS of PU's transmission. Hence, a spectrum-sharing wireless communication setup is considered here, where SU can transmit over the shared wireless channel if it complies with the QoS requirements of PU.

SUs are split into two groups: SU_1 is the IoT device of interest, and SU_2 represents the remaining IoT devices in the system. SU_1 and SU_2 denote the controller user and the opponent user, respectively. The main objective is to develop a transmission strategy for SU_1 that enables this device to communicate over the channel and also meet the QoS constraints of PU regardless of the transmission activities of SU_2. The transmission strategy of SU_1 should also allow the remaining SU_2 nodes in the industrial WSN to communicate their measurements using spectrum-sharing cognitive communication without coordination while meeting the QoS requirement of PU.

To facilitate a technical discussion, consider a setting where there are two or more SUs in the communication system that want to concurrently share the channel with PU and transmit data to RU. This concurrent secondary transmission leads to interference on the primary signal received at RU. Figure 7.2 shows the communication system setting for the primary user and two secondary users.

First, let P_p and R_p denote PU's transmission power and data rate over the channel, respectively, and let P_s denote SU's transmission power. Then, P_{s_1} means the transmission power of SU_1, and P_{s_2} is the transmission power of SU_2. The received signals at RU are corrupted with an additive white Gaussian noise with a mean of zero and a variance of σ^2. Next, assume the wireless channels between PU, SU_1, and SU_2 with RU experience independent and identically distributed (i.i.d.) block fading with a Rayleigh distribution for the channel gains. Thus, the channel power

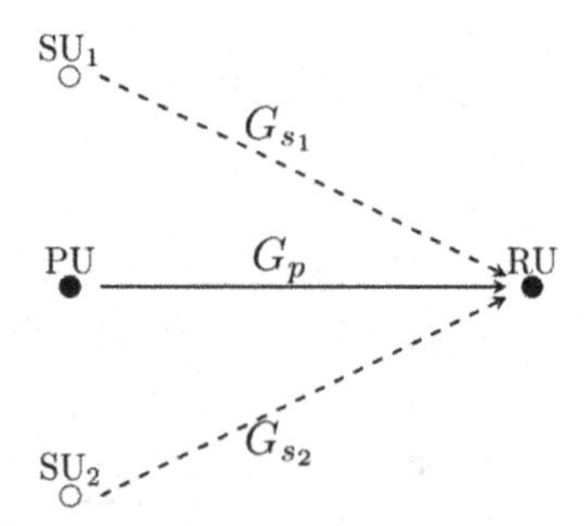

Figure 7.2 Model of the considered system showing the channels between the IoT devices and the receiver unit.

gains between PU, SU_1, and SU_2 with RU are denoted as G_p, G_{s_1}, and G_{s_2}, each having an exponential distribution.

Now, let the QoS constraint that SUs have to adhere to be the outage probability of the received PU's signal at RU. Thus, to transmit over the shared wireless channel, both SU_1 and SU_2 must not increase the primary outage probability beyond a QoS limit named ζ. Let the SINR of the received PU's signal at RU be denoted as γ_p, and let $\mathbb{P}\{.\}$ be the probability operator. Consequently, the QoS constraint on the primary outage probability at RU using the IoT transmission is defined as $\mathbb{P}\{\log_2(1+\gamma_p) \leq R_p\} \leq \zeta$ [15].

7.3.3 PRIMARY OUTAGE PROBABILITY

Now, we investigate the impact of the concurrent secondary transmissions on the primary outage probability. When there is no secondary transmission, the signal-to-noise ratio (SNR) of the received PU's signal at RU is expressed as $\gamma_p = \frac{P_p G_p}{\sigma^2}$. Then, the primary outage probability, denoted for this case as ζ_0, is calculated from $\zeta_0 = \mathbb{P}\{\log_2(1+\gamma_p) \leq R_p\}$, or $\zeta_0 = \mathbb{P}\{\gamma_p \leq 2^{R_p} - 1\}$. Consequently:

$$\zeta_0 = 1 - \exp\left(\frac{1 - 2^{R_p}}{P_P/\sigma^2}\right). \tag{7.1}$$

Next, the outage probability under spectrum-sharing secondary communications is considered. When the shared channel is being accessed by SU along with PU, the SINR of PU's signal becomes $\gamma_p = \frac{P_p G_p}{P_s G_s + \sigma^2}$ for a secondary transmission power of P_s with channel power gain of G_s. Let f_{γ_p} be the probability density function (PDF) of γ_p; then:

$$f_{\gamma_p}(z) = \int_{\sigma^2}^{\infty} \frac{1}{P_p} \frac{1}{P_s} y \exp\left(-\frac{yz}{P_p}\right) \exp\left(\frac{\sigma^2 - y}{P_s}\right) dy. \tag{7.2}$$

This leads to:

$$f_{\gamma_p}(z) = \frac{P_p}{P_s} \exp\left(-\frac{\sigma^2}{P_p} z\right) \frac{1}{(z+\frac{P_p}{P_s})^2} \left(\frac{\sigma^2}{P_p}(z+\frac{P_p}{P_s})+1 \right). \tag{7.3}$$

The outage probability of PU is then derived from $\zeta_s = \mathbb{P}\{\gamma_p \leq 2^{R_p} - 1\}$, leading to:

$$\zeta_s = 1 + \frac{\zeta_0 - 1}{1 + \frac{P_s}{P_p}(2^{R_p} - 1)}. \tag{7.4}$$

Comparing Eqs. (7.1) and (7.4), it is observed that PU experiences higher outage probability compared to the case when there is no secondary transmission because $\zeta_0 < \zeta_s$ for $P_s > 0$. Thus, when SUs concurrently transmit their data over the shared channel along with PU, they need to comply with $\zeta_s \leq \zeta$. As PU allows uncoordinated secondary transmissions, it is understood here that $\zeta_0 < \zeta_s \leq \zeta$, where ζ was defined before as the primary QoS threshold.

7.4 ZERO-DETERMINANT STRATEGIES

In this section, we describe a game-theoretic strategy for iterated games, termed *zero-determinant*, in order to devise an uncoordinated channel access for the controller user (i.e., SU_1) to meet the outage probability constraints of PU regardless of the transmission activity of other opponent users (i.e., SU_2). The zero-determinant game approach fits the nature of repeated uncoordinated transmissions of IoT devices in an industrial WSN.

A generic payoff matrix of a 2×2 iterated game is depicted in Figure 7.3. Two players are in the game: User 1 (also called the controller user) is the row player in the payoff matrix, and User 2 (also called the adversary user) is the column player in the matrix. A player in this iterated game chooses one from two actions $\{1,2\}$ at each round of the game. The game actions of User 1 and User 2 are denoted as n_1 and n_2, respectively. Let $n_1 = 1$ indicate an active User 1 in a given round of the game; also, let $n_1 = 2$ mean an idle User 1 in a game round. Similar notation applies to the User 2: $n_2 = 1$ and $n_2 = 2$ mean an active adversary and an idle adversary, respectively. Further, for $j,k \in \{1,2\}$, the value $X_{j,k}$ in Figure 7.3 represents the payoff of a game round when User 1 chooses action $n_1 = j$ and User 2 chooses action $n_2 = k$.

User 2 / User 1	$n_2 = 1$ (Active)	$n_2 = 2$ (Idle)
$n_1 = 1$ (Active)	$X_{1,1}$	$X_{1,2}$
$n_1 = 2$ (Idle)	$X_{2,1}$	$X_{2,2}$

Figure 7.3 Generic payoff matrix of a 2×2 iterated zero-determinant game.

In addition, when the same payoff matrices and the same users' actions are repeated in the iterated game, any history of user actions outside what is shared between the game players can be ignored as shown in Ref. [29]. Then, if this is the case, a Markov chain approach can be utilized to model this iterated game. Thus, let $\boldsymbol{n}(t) = (n_1, n_2)$ represent the state of the iterated game at round t, and let $\boldsymbol{S} = \{(1,1),(1,2),(2,1),(2,2)\}$ denote the state space of the game at round t. In addition, let $\boldsymbol{k} = (k_1, k_2)$, where $k_1, k_2 \in \{1,2\}$, be the probability that User 1 takes action 1 in round $t+1$ given that User 1 took action k_1 and User 2 took action k_2 in round t is then expressed as

$$p_1^{\boldsymbol{k}} = \mathbb{P}(n_1(t+1) = 1 \mid \boldsymbol{n}(t) = \boldsymbol{k}), \forall \boldsymbol{k} \in \boldsymbol{S}. \tag{7.5}$$

For User 2, the probability that $n_2 = 1$ in round $t+1$ if User 1 took action k_1 and User 2 took action k_2 in round t is calculated as:

$$p_2^{\boldsymbol{k}} = \mathbb{P}(n_2(t+1) = 1 \mid \boldsymbol{n}(t) = \boldsymbol{k}), \forall \boldsymbol{k} \in \boldsymbol{S}. \tag{7.6}$$

Let the stationary probability distribution of User 1 taking action j and User 2 taking action k be termed $\pi_{j,k}$, $\forall j,k \in \{1,2\}$. Then, the Markov chain that can be employed to model this iterated game has $\pi = [\pi_{1,1}, \pi_{1,2}, \pi_{2,1}, \pi_{2,2}]^T$ as a stationary distribution. Let $\hat{\boldsymbol{X}} = [X_{1,1}, X_{1,2}, X_{2,1}, X_{2,2}]^T$; then, the long-term average payoff of the iterated game, denoted as u_X, can be found from Ref. [2]

$$u_X = \pi^T \hat{\boldsymbol{X}}. \tag{7.7}$$

Let a and b be two arbitrary non-zero real numbers, and let the action probabilities be expressed as:

$$a\hat{\boldsymbol{X}} + b = \left[-1 + p_1^{1,1}, -1 + p_1^{1,2}, p_1^{2,1}, p_1^{2,2}\right]^T. \tag{7.8}$$

It is shown in Ref. [29] that if $p_1^{\boldsymbol{k}}$'s are chosen following Eq. (7.8), then User 1 can fix u_X of the iterated game regardless of the game actions of User 2 if and only if the minimum value of one row in the payoff matrix of Figure 7.3 exceeds the maximum value of the other row of the matrix. Furthermore, User 1 can fix u_X to any value in the range between those minimum and the maximum values [2].

Let $p_1^{1,1}$, $p_1^{1,2}$, $p_1^{2,1}$, and $p_1^{2,2}$ denote the probability that User 1 is active in the current play round given that $n_1 = 1$ & $n_2 = 1$, $n_1 = 1$ & $n_2 = 2$, $n_1 = 2$ & $n_2 = 1$, and $n_1 = 2$ & $n_2 = 2$ in the previous interval, respectively. To achieve a specific value of u_X, User 1 takes an *Active* action in any game round with probabilities of Ref. [2]:

$$\begin{aligned}
p_1^{1,1} &= 1 + \left(1 - \frac{X_{1,1}}{u_X}\right) b \\
p_1^{1,2} &= 1 + \left(1 - \frac{X_{1,2}}{u_X}\right) b \\
p_1^{2,1} &= \left(1 - \frac{X_{2,1}}{u_X}\right) b \\
p_1^{2,2} &= \left(1 - \frac{X_{2,2}}{u_X}\right) b.
\end{aligned} \tag{7.9}$$

Further, b in Eq. (7.8) needs to have a valid value in the range of Ref. [2]:

$$0 < b \leq \min\left(\frac{-1}{1-\frac{X_{1,\max}}{u_X}}, \frac{1}{1-\frac{X_{2,\min}}{u_X}}\right). \tag{7.10}$$

7.5 GAME-THEORETIC STRATEGY FOR IoT TRANSMISSION

We apply the game-theoretic approach to address the IoT channel access problem in this section. We employ the zero-determinant strategy of repeated games to formulate an uncoordinated channel access strategy for industrial IoT devices (SUs) while meeting the QoS requirement.

7.5.1 UNCOORDINATED TRANSMISSION STRATEGY

Recall that the IoT device of interest wants to transmit the data over the shared channel without coordinating its transmission activities with other IoT devices while meeting the primary outage constraint. Consequently, we utilize the value of the outage probability ζ_s to represent the *payoff* of the game at the end of each interval ΔT. Thus, representing the role of SU_1, User 1 wants to communicate over the wireless channel; also, User 1 wants to satisfy the QoS requirements of PU regardless of the channel access activity of the other SUs in the communication environment. As a starting point, the collective action of the rest of SUs in the system is modeled as SU_2 (i.e., the opponent user). In the following discussion, we focus on the actions of SU_1 as the controller user.

Let $\boldsymbol{X} = \left[X_{j,k}\right]$ denote the payoff matrix of PU during interval ΔT. As previously described in Section 7.3, the values in $\boldsymbol{X}$ represent the primary outage probabilities as shown in Figure 7.4. In this notation, SU is *Active* means that SU transmits over the channel during interval ΔT; similarly, an *Idle* SU indicates there is no secondary transmission during the interval. Furthermore, let $X_{j,\max}$ and $X_{j,\min}$ stand for the maximum and minimum values of row j in $\boldsymbol{X}$, respectively. From Figure 7.4, these values are found as:

SU_2 / SU_1	$n_2 = 1$ (Active)	$n_2 = 2$ (Idle)
$n_1 = 1$ (Active)	$1 + \frac{(\zeta_0 - 1)P_p}{P_p + (P_{s_1} + P_{s_2})(2^{R_p} - 1)}$	$1 + \frac{(\zeta_0 - 1)P_p}{P_p + P_{s_1}(2^{R_p} - 1)}$
$n_1 = 2$ (Idle)	$1 + \frac{(\zeta_0 - 1)P_p}{P_p + P_{s_2}(2^{R_p} - 1)}$	ζ_0

Figure 7.4 Payoff matrix of the primary user of the system showing the outage probabilities.

$$
\begin{aligned}
X_{1,\max} &= X_{1,1} = 1 + \frac{(\zeta_0-1)P_p}{P_p+(P_{s_1}+P_{s_2})(2^{R_p}-1)} \\
X_{1,\min} &= X_{1,2} = 1 + \frac{(\zeta_0-1)P_p}{P_p+P_{s_1}(2^{R_p}-1)} \\
X_{2,\max} &= X_{2,1} = 1 + \frac{(\zeta_0-1)P_p}{P_p+P_{s_2}(2^{R_p}-1)} \\
X_{2,\min} &= X_{2,2} = \zeta_0 .
\end{aligned} \tag{7.11}
$$

Consider the case of an empowered User 1 that has higher transmission power than the opponent IoT device (i.e., $P_{s_1} > P_{s_2}$); then, it is observed that $X_{2,\min} < X_{2,\max} < X_{1,\min} < X_{1,\max}$. Thus, $X_{2,\max} < X_{1,\min}$ is satisfied. Consequently, the long-term average payoff (u_X) that can be achieved using the zero-determinant strategies by SU_1 can be in the range $[X_{2,1}, X_{1,2}]$, or

$$
u_X \in 1 + \left[\frac{(\zeta_0-1)P_p}{P_p+P_{s_2}(2^{R_p}-1)}, \frac{(\zeta_0-1)P_p}{P_p+P_{s_1}(2^{R_p}-1)} \right]. \tag{7.12}
$$

Consider the case when u_X has to be a specific value in the valid range, and let $0 \leq \alpha \leq 1$ be termed the *persistence factor*. Then, u_X can be parameterized using:

$$
\begin{aligned}
u_X &= X_{2,1} + \alpha(X_{1,2} - X_{2,1}) \\
&= 1 + \frac{(\zeta_0-1)\alpha P_p}{P_p+P_{s_1}(2^{R_p}-1)} + \frac{(\zeta_0-1)(1-\alpha)P_p}{P_p+P_{s_2}(2^{R_p}-1)} .
\end{aligned} \tag{7.13}
$$

In this notation, a higher value of α shifts u_X closer to its upper limit ($X_{1,2}$), implying that SU_1 is more aggressive in transmitting over the wireless channel while satisfying the QoS constraint.

Further, following the findings of Ref. [2], b has a range of values from Eq. (7.8) as:

$$
0 < b \leq \min\left(\frac{-u_X}{u_X - X_{1,\max}}, \frac{u_X}{u_X - X_{2,\min}} \right). \tag{7.14}
$$

Define:

$$
b_{\max} = \begin{cases} \frac{u_X}{X_{1,\max}-u_X} & u_X \leq \frac{X_{1,\max}+X_{2,\min}}{2} \\ \frac{u_X}{u_X-X_{2,\min}} & u_X > \frac{X_{1,\max}+X_{2,\min}}{2} \end{cases} . \tag{7.15}
$$

Let $0 < \beta \leq 1$ be denoted as the *steering factor*; then, the value of b that SU_1 will be using in the transmission strategy can then be expressed as:

$$
b = \beta b_{\max} . \tag{7.16}
$$

Similarly, higher values of β indicate that SU_1 is more probable to be *Active* in the current transmission interval (i.e., SU_1 transmits data to RU) if it was idle in the previous one, and SU_1 is less probable to transmit over the channel in the current transmission ΔT if the IoT device was active in the previous ΔT.

For $j,k \in \{1,2\}$, let the transmission status of SU_1 and SU_2 in the previous ΔT be j and k, respectively. Also, assume j and k are known to SU_1. Thus, $p_1^{j,k}$ means the probability that SU_1 is active on the channel in the current ΔT given that it knows that $n_1 = j$ and $n_2 = k$ in the previous ΔT. Then, from Eqs. (7.9) and (7.13), the probabilities of SU_1 transmitting over the channel in a ΔT are expressed as:

$$\begin{aligned} p_1^{1,1} &= 1 + \frac{X_{2,1}+\alpha(X_{1,2}-X_{2,1})-X_{1,1}}{X_{2,1}+\alpha(X_{1,2}-X_{2,1})} b \\ p_1^{1,2} &= 1 + \frac{X_{2,1}-X_{1,2}}{X_{2,1}+\alpha(X_{1,2}-X_{2,1})}(1-\alpha)b \\ p_1^{2,1} &= \frac{X_{1,2}-X_{2,1}}{X_{2,1}+\alpha(X_{1,2}-X_{2,1})}\alpha b \\ p_1^{2,2} &= \frac{X_{2,1}+\alpha(X_{1,2}-X_{2,1})-X_{2,2}}{X_{2,1}+\alpha(X_{1,2}-X_{2,1})} b. \end{aligned} \tag{7.17}$$

Given this development, the controller user (SU_1) can adopt a transmission strategy that takes into consideration the most-recent transmission actions of the opponent and controller users and still satisfies the QoS constraint of PU. Algorithm 3 depicts the zero-determinant transmission strategy that SU_1 employs to transmit over the wireless channel while maintaining u_X as a long-term average payoff and meeting the QoS requirement ζ of PU.

Algorithm 3 IoT Transmission Strategy

Collect values of ζ, R_p, P_p, σ^2, and P_{s_2}.
Determine value of transmission power P_{s_1}.
Calculate payoff matrix X.
Determine values of parameters α and β.
Calculate long-term average payoff u_X that satisfies outage probability requirement ζ.
Calculate p_1 for channel access probabilities.
Set $j = 2$ as a starting *Idle* status in the previous ΔT.
while TRUE **do**
  Find transmission status of SU_2 in the previous interval ΔT.
  Determine value of $k \in \{1,2\}$.
  Calculate value of $p_1^{j,k}$.
  Generate a random number $0 \leq r \leq 1$.
  if $p_1^{j,k} \geq r$ **then**
    Channel access: SU_1 transmits data over the channel with transmission power P_{s_1}.
    Assign $j = 1$.
  else
    Channel idle: SU_1 does not transmit over the channel.
    Assign $j = 2$.
  end if
  if SU_1 has no more data to transmit, **then**
    Exit *IoT Transmission Strategy.*
  end if
end while

7.5.2 SPECIAL CASES

We now focus on the study of few special cases of Eq. (7.17) to gain more insights on the proposed IoT device's transmission strategy. The cases highlight the difference in transmission strategy when the controller user adopts competitive versus noncompetitive strategies; in addition, the we consider the practical case when the system users utilize the same transmission power.

7.5.2.1 Case of Competitive Strategy

Consider the case when SU_1 adopts a more aggressive transmission strategy by choosing $u_X > \frac{X_{1,\max}+X_{2,\min}}{2}$. In this case, $b_{\max} = \frac{u_X}{u_X - X_{2,\min}}$ in Eq. (7.15); as a result, the transmission probabilities become:

$$\begin{aligned} p_1^{1,1} &= 1+\beta\frac{(1-\alpha)X_{2,1}+\alpha X_{1,2}-X_{1,1}}{(1-\alpha)X_{2,1}+\alpha X_{1,2}-X_{2,2}} \\ p_1^{1,2} &= 1-\beta(1-\alpha)\frac{X_{2,1}-X_{1,2}}{(1-\alpha)X_{2,1}+\alpha X_{1,2}-X_{2,2}} \\ p_1^{2,1} &= \beta\alpha\frac{X_{1,2}-X_{2,1}}{(1-\alpha)X_{2,1}+\alpha X_{1,2}-X_{2,2}} \\ p_1^{2,2} &= \beta. \end{aligned} \tag{7.18}$$

Further, $u_X = X_{2,1}$ when $\alpha = 1$. Consequently, the transmission probabilities of SU_1 when $\alpha = 1$ become:

$$\begin{aligned} p_1^{1,1} &= 1-\beta\frac{P_{s_2}}{P_{s_1}}\frac{P_p}{P_p+(P_{s_1}+P_{s_2})(2^{R_p}-1)} \\ p_1^{1,2} &= 1 \\ p_1^{2,1} &= \beta\frac{P_{s_1}-P_{s_2}}{P_{s_1}}\frac{P_p}{P_p+P_{s_2}(2^{R_p}-1)} \\ p_1^{2,2} &= \beta. \end{aligned} \tag{7.19}$$

The above result signifies that SU_1's competitive strategy leads to certain probability of transmission if the opponent user was idle during the previous interval. Furthermore, when both users were idle in the previous interval, the controller user will transmit over the channel during the current interval with a probability of β; as was described earlier, a higher value of β leads to SU_1 being more probable to be active in the current transmission interval if it was idle in the previous interval.

7.5.2.2 Case of Noncompetitive Strategy

When SU_1 adopts a less competitive strategy by choosing $u_X \leq \frac{X_{1,\max}+X_{2,\min}}{2}$, Eq. (7.15) simplifies to $b_{\max} = \frac{u_X}{X_{1,\max}-u_X}$. The transmission probabilities become:

$$\begin{aligned} p_1^{1,1} &= 1-\beta \\ p_1^{1,2} &= 1-\beta\frac{u_X-X_{1,2}}{u_X-X_{1,1}} \\ p_1^{2,1} &= -\beta\frac{u_X-X_{2,1}}{u_X-X_{1,1}} \\ p_1^{2,2} &= -\beta\frac{u_X-X_{2,2}}{u_X-X_{1,1}}. \end{aligned} \tag{7.20}$$

Similarly, when $\alpha = 0$, SU_1 settles for the lower bound of the payoff range leading to $u_X = X_{2,1}$. This results in the following transmission probabilities:

$$\begin{aligned} p_1^{1,1} &= 1-\beta \\ p_1^{1,2} &= 1-\beta\frac{X_{2,1}-X_{1,2}}{X_{2,1}-X_{1,1}} \\ p_1^{2,1} &= 0 \\ p_1^{2,2} &= -\beta\frac{X_{2,1}-X_{2,2}}{X_{2,1}-X_{1,1}}. \end{aligned} \tag{7.21}$$

The noncompetitive strategy of SU_1 leads to the case where the controller user does not utilize the channel if during the previous interval it was idle and the opponent user was active; this strategy achieves the lowest payoff possible.

7.5.2.3 Case of Equal Transmission Power

Consider the case when SUs have equal transmission powers leading to $P_{s_1} = P_{s_2} = P_s$. For this case, the long-term average payoff becomes:

$$\begin{aligned} u_X = X_{1,2} = X_{2,1} &= 1+\frac{(\zeta_0-1)P_p}{P_p+P_s(2^{R_p}-1)} \\ X_{1,1} &= 1+\frac{(\zeta_0-1)P_p}{P_p+2P_s(2^{R_p}-1)}. \end{aligned} \tag{7.22}$$

As such, SU_1 can access the wireless channel with transmission probabilities of:

$$\begin{aligned} p_1^{1,1} &= \begin{cases} 1-\beta\frac{P_p}{P_p+2P_s(2^{R_p}-1)} & b_{\max} = \frac{u_X}{u_X-X_{2,\min}} \\ 1-\beta & b_{\max} = \frac{u_X}{X_{1,\max}-u_X} \end{cases} \\ p_1^{1,2} &= 1 \\ p_1^{2,1} &= 0 \\ p_1^{2,2} &= \begin{cases} \beta & b_{\max} = \frac{u_X}{u_X-X_{2,\min}} \\ \beta\frac{P_p}{P_p+2P_s(2^{R_p}-1)} & b_{\max} = \frac{u_X}{X_{1,\max}-u_X} \end{cases}. \end{aligned} \tag{7.23}$$

This situation resembles the practical case of IoT devices with similar transmission settings. The results above indicates that the controller user will be certainly active in the channel if during the previous interval it was active while the opponent user was idle. On the other hand, SU_1 will not be transmitting over the channel if during the past interval it was idle and SU_2 was active.

7.5.3 PERFORMANCE ANALYSIS

The probability that SU_1 takes an active action (i.e., p_1) is defined in Eq. (7.5); also, p_2, the probability that opponent user takes an active action, is defined in Eq. (7.6). Let the state transition matrix of the Markov chain be denoted $\boldsymbol{M}$ and is defined as:

$$M = \begin{bmatrix} p_1^{1,1}p_2^{1,1} & p_1^{1,1}(1-p_2^{1,1}) & (1-p_1^{1,1})p_2^{1,1} & (1-p_1^{1,1})(1-p_2^{1,1}) \\ p_1^{1,2}p_2^{2,1} & p_1^{1,2}(1-p_2^{2,1}) & (1-p_1^{1,2})p_2^{2,1} & (1-p_1^{1,2})(1-p_2^{2,1}) \\ p_1^{2,1}p_2^{1,2} & p_1^{2,1}(1-p_2^{1,2}) & (1-p_1^{2,1})p_2^{1,2} & (1-p_1^{2,1})(1-p_2^{1,2}) \\ p_1^{2,2}p_2^{2,2} & p_1^{2,2}(1-p_2^{2,2}) & (1-p_1^{2,2})p_2^{2,2} & (1-p_1^{2,2})(1-p_2^{2,2}) \end{bmatrix}. \quad (7.24)$$

Let π be the stationary distribution calculated using [2,29]:

$$\pi^T = [\pi_{1,1}, \pi_{1,2}, \pi_{2,1}, \pi_{2,2}] = \pi^T M. \quad (7.25)$$

As previously described, $\pi_{j,k}$, $\forall j,k \in \{1,2\}$, denotes the stationary probability distribution of SU_1 taking action j while the opponent user taking action k. Since $\sum \pi = 1$, the stationary distribution π can be calculated from the normalized left eigenvector of the transition matrix M that has a corresponding eigenvalue of 1. Consequently, SU_1's transmission probability is $\pi_{1,1} + \pi_{1,2}$.

Define the primary channel's capacity vector as:

$$C_p = \log_2\left(1 + P_P G_p / \begin{bmatrix} P_{s_1}G_{s_1} + P_{s_2}G_{s_2} + \sigma^2 \\ P_{s_1}G_{s_1} + \sigma^2 \\ P_{s_2}G_{s_2} + \sigma^2 \\ \sigma^2 \end{bmatrix}\right). \quad (7.26)$$

Then, the mean capacity of PU's channel is found as $\overline{C_p} = \pi^T \cdot C_p$. Similarly, let the channel's capacity vector of the controller user be defined as:

$$C_s = \log_2\left(1 + P_{s_1}G_{s_1} / \begin{bmatrix} P_P G_p + P_{s_2}G_{s_2} + \sigma^2 \\ P_P G_p + \sigma^2 \end{bmatrix}\right). \quad (7.27)$$

The mean channel capacity of the controller user is then calculated using $\overline{C_s} = \frac{1}{\pi_{1,1}+\pi_{1,2}} [\pi_{1,1}, \pi_{1,2}]\, C_s$.

Given the developed zero-determinant transmission strategy depicted in Algorithm 3, the controller user can choose to be an active user of the wireless channel or idle to conserve resources. Moreover, the controller user in the game does not have to know the entire history of the opponent user's transmission activity in order to utilize the IoT transmission strategy. User 1 can take a transmission action in the current interval based only on the transmission state of the opponent user in the previous interval.

Recall that the long-term average payoff achieved using the proposed zero-determinant strategy lies in $[X_{1,1}, X_{1,2}]$. Further, if SU_1 chooses to transmit over the shared wireless channel in all intervals (i.e., not following Algorithm 3), the long-term average payoff will be in the range of $[X_{2,1}, X_{1,2}]$, which is higher than that achieved by utilizing the proposed transmission strategy. Nevertheless, the proposed transmission strategy gives the secondary users a guarantee of meeting the QoS requirement of PU while satisfying any constraints on channel-access costs or data availability.

7.6 EXTENSION TO MULTIPLE USERS

The analysis presented for two players is next extended as a repeated game with multiple players. This is necessary to examine the scalability and generality of the presented model. Let $N \geq 2$ be the number of game players with $\{1,\ldots,N\}$ being the index of the players. Let $\boldsymbol{n}(t) = [n_1(t),\ldots,n_N(t)]$ represent the state of the repeated game at round t, where $n_i(t) \in \{1,2\}$ describes the *Active* or *Idle* binary actions $\forall t, i \in \{1,\ldots,N\}$. A multi-dimensional Markov chain can be used to describe the process $\{\boldsymbol{n}(t) : t = 0,1,\ldots\}$, and the state transition matrix, $\boldsymbol{M}$, can be presented using a $2^N \times 2^N$ matrix.

Similar to the two-player game, a player i, $\forall i = \{1,\ldots,N\}$, in an N-player game takes a specific action in a given round with a probability that depends on the *Active* or *Idle* actions of the players in the previous round of the game. Further, let $\boldsymbol{X}_i^{\boldsymbol{k}}$ refer to the payoff value of player i in the current round if the state of the game is $\boldsymbol{k}$ at the previous round, and let $p_i^{\boldsymbol{k}}$ be the probability that player i takes action 1 in a given game round if the game is in state $\boldsymbol{k}$ in the previous one. Also, define $\boldsymbol{X}_{i,\min}^{k} = \min(\boldsymbol{X}_i^{\boldsymbol{k}} : n_i = k)$ and $\boldsymbol{X}_{i,\max}^{k} = \max(\boldsymbol{X}_i^{\boldsymbol{k}} : n_i = k)$, for $k \in \{1,2\}$, as the minimum and maximum payoffs of player i when taking action k, respectively.

Following the results of Ref. [2], player i can fix the long-term average of its game payoff regardless of the game actions of the other players if $k_{i,\max}, k_{i,\min} \in \{1,2\}$ exist such that $\boldsymbol{X}_{i,\max}^{k_{\max}} \leq \boldsymbol{X}_{i,\min}^{k_{\min}}$. In this case, the long-term average payoff of player i, termed u_i, can be any value in the interval $[\boldsymbol{X}_{i,\max}^{k_{\max}}, \boldsymbol{X}_{i,\min}^{k_{\min}}]$, and this long-term payoff can be achieved using the strategy of:

$$p_i^{\boldsymbol{k}} = 1 + \frac{b_i}{u_i}(u_i - \boldsymbol{X}_i^{\boldsymbol{k}}) \tag{7.28}$$

as the probability of choosing action 1 when the state of the game is $\boldsymbol{k}$ where b_i depends on the value of $k_{i,\max}$ [2].

Given the N SUs in the communication system and for SU_1 being the controller user, SU_1 takes actions whether to transmit over the wireless channel following the zero-determinant strategy described above. To meet the QoS constraints of PU, SU_1 conducts a zero-determinant transmission strategy as follows:

- Calculate the $N \times N$ payoff matrix of PU given the outage probabilities as demonstrated in Figure 7.4
- Verify if $\boldsymbol{X}_{1,\max}^{k_{\max}} \leq \boldsymbol{X}_{1,\min}^{k_{\min}}$
- Define the long-term target of PU's outage probability in the range $[\boldsymbol{X}_{1,\max}^{k_{\max}}, \boldsymbol{X}_{1,\min}^{k_{\min}}]$ as

$$u_1 = \boldsymbol{X}_{1,\max}^{k_{\max}} + \alpha_1 \left(\boldsymbol{X}_{1,\min}^{k_{\min}} - \boldsymbol{X}_{1,\max}^{k_{\max}} \right) \tag{7.29}$$

- Select b_1 given the value of $k_{1,\max}$
- For each transmission interval ΔT:
  - Determine the previous ΔT's transmission state, $\boldsymbol{k}$

– Determine the previous ΔT's game payoff of SU_1, X_1^k
– Transmit over the wireless channel with probability (7.28)

$$p_1^k = 1 + \frac{b_1}{u_1}\left(u_1 - X_1^k\right). \tag{7.30}$$

As evident from the above description, the N-user case is a natural extension of the 2-player case detailed in Algorithm 3.

7.7 NUMERICAL RESULTS

To validate the proposed approach, we investigate key metrics and relations through two sets of simulations results: (i) numerical results of the analytical solution for different sets of parameters and (ii) simulation results to verify the analytical approach.

The primary user's outage probability in Eqs. (7.1) and (7.4) is numerically investigated in Figure 7.5. Figure 7.5a shows the case when there is no secondary transmission, and it confirms that $\zeta_0 = 1 - \exp\left(\frac{1-2^{R_p}}{P_P/\sigma^2}\right)$ increases as SNR decreases and/or when R_p increases. Similarly, during secondary transmission, the primary outage probability in Figure 7.5b and c increases when the secondary interference and/or the noise power increases.

The relation between p_1 (SU_1's *Active* probabilities) and α (persistence factor) and β (steering factor) is investigated in Figure 7.6. For this figure, $\frac{P_p}{P_{s_2}} = 11.25$ dB, $\frac{P_p}{P_{s_1}} = 10$ dB, $\frac{P_p}{\sigma^2} = 20$ dB, and $R_p = 1$ bit/sec/Hz are used. We observe that α has a direct impact on p_1 as defined in Eq. (7.17). However, the impact of the value of β on p_1 is affected by the specific value of α as demonstrated in Eqs. (7.18)–(7.23).

We then simulate a communication environment that has two IoT devices modeled as SUs and a PU of the channel as illustrated in Figure 7.2. For this environment, PU's outage probability is not to exceed $\zeta = 12.5\%$; also, $\frac{P_p}{P_{s_2}} = 11$ dB, $\frac{P_p}{P_{s_1}} = 10$ dB, $\frac{P_p}{\sigma^2} = 15$ dB, and $R_p = 1$ bit/sec/Hz. Also, SU_2 (the opponent user) transmits data randomly over the shared wireless channel with probability of 50%. In addition, SU_1 (the controller user) adopts the transmission strategy in Algorithm 3 with $\alpha = 0.5$ and $\beta = 0.5$. Adopting the zero-determinant by SU_1 leads to developing the payoff matrix found in Figure 7.7. Also, SU_1 chooses $u_X = 11.1\%$, which is less than the maximum PU outage constraint of $\zeta = 12.5\%$. From Eq. (7.17), the transmission probabilities for SU_1 become

$$\begin{array}{ll} p_1^{1,1} = 0.575 & p_1^{1,2} = 0.947 \\ p_1^{2,1} = 0.053 & p_1^{2,2} = 0.500. \end{array} \tag{7.31}$$

With $u_X < \zeta$, using the IoT transmission strategy, SU_1 is able to meet the QoS constraints of PU regardless of the transmission activity of SU_2 (the opponent user).

Sample long-term performance metrics are captured in Figure 7.8 of this simulation environment. Figure 7.8a illustrates the attained value of u_X over time, which reflects the resultant outage probability of PU due to the IoT transmission activity;

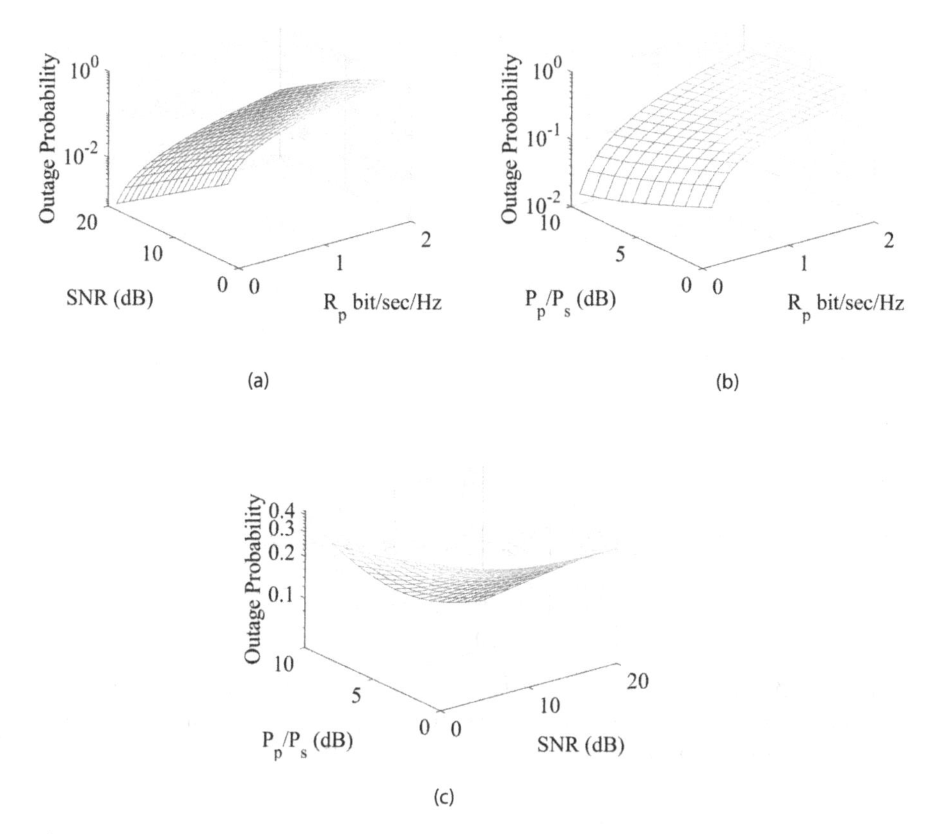

Figure 7.5 Outage probability of primary user of the system versus different communication parameters. (a) Outage probability of the primary user vs. SNR and primary transmission rate. (b) Outage probability of the primary user for SNR = P_p/σ^2 = 10 dB. (c) Outage probability of the primary user for $R_p = 0.5$ bit/sec/Hz.

the results here show how SU_1's transmission strategy meets the QoS requirement regardless of the transmission activity of SU_2. Figure 7.8b shows that the probability of SU_1 being *Active* in a game round is about 53.8% which is equal to $(p_1^{1,1} + p_1^{2,2})/2$. Figure 7.8c and d show the long-term channel capacity for the primary user and the IoT of interest, respectively. These figures demonstrate the advantage of the proposed strategy where the average channel capacity of PU is shown to be better than the case when both IoT devices simultaneously transmit over the wireless channel. Also, the proposed algorithm has the advantage of providing uncoordinated transmission over the channel enabled by local information gathered by the users of the system.

Finally, Figure 7.9 considers the multiple-user case and illustrates the effect of increasing the number of IoT devices on the performance metrics. The parameters of simulation environment are $R_p = 1$ bit/sec/Hz, $\frac{P_p}{\sigma^2} = 20$ dB; $\frac{P_p}{P_{s_1}} = 10$ dB, $\frac{P_p}{P_{s_i}} = 20$ dB $\forall i \in \{2,\ldots,N\}$, $\alpha = 0.5$, and $\beta = 0.5$. One observation here is that the communi-

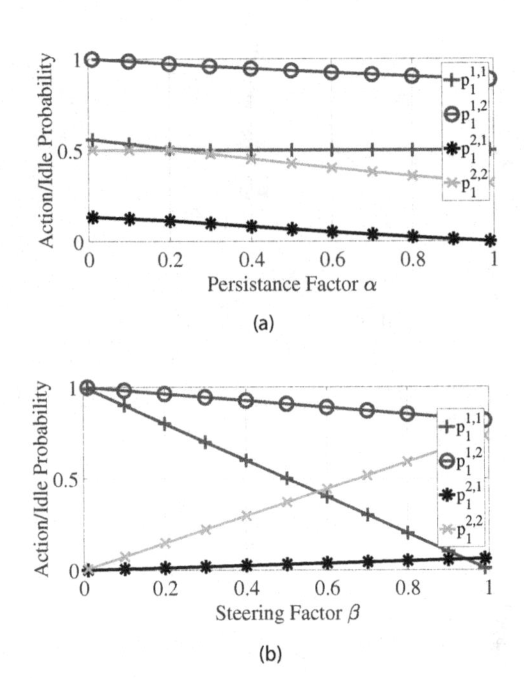

Figure 7.6 *Active & Idle* probabilities of the primary user of the system versus α (persistence factor) and β (steering factor). (a) *Active & Idle* probabilities of the primary user for $\beta = 0.5$. (b) *Active & Idle* probabilities of the primary user for $\alpha = 0.75$.

SU_1 \ SU_2	Active	Idle
Active	17.9%	11.9%
Idle	10.2%	3.1%

Figure 7.7 Calculated payoff matrix for the primary user for the four combinations of *Active & Idle*.

cation system experiences *soft limit* behavior where increasing the number of IoT devices in the system gracefully deteriorates the performance metrics of the system even though the channel access is uncoordinated.

7.8 DISCUSSIONS AND CONCLUSIONS

In this chapter, we present a framework to investigate dynamical interaction between the IoT devices when accessing a shared wireless channel that is licensed to a primary user. In contrast to coordinated channel access schemes where the "*winner takes all*", we employ valid constraints on uncoordinated transmission cycles and data availability to propose an uncoordinated IoT transmission strategy to optimize information availability.

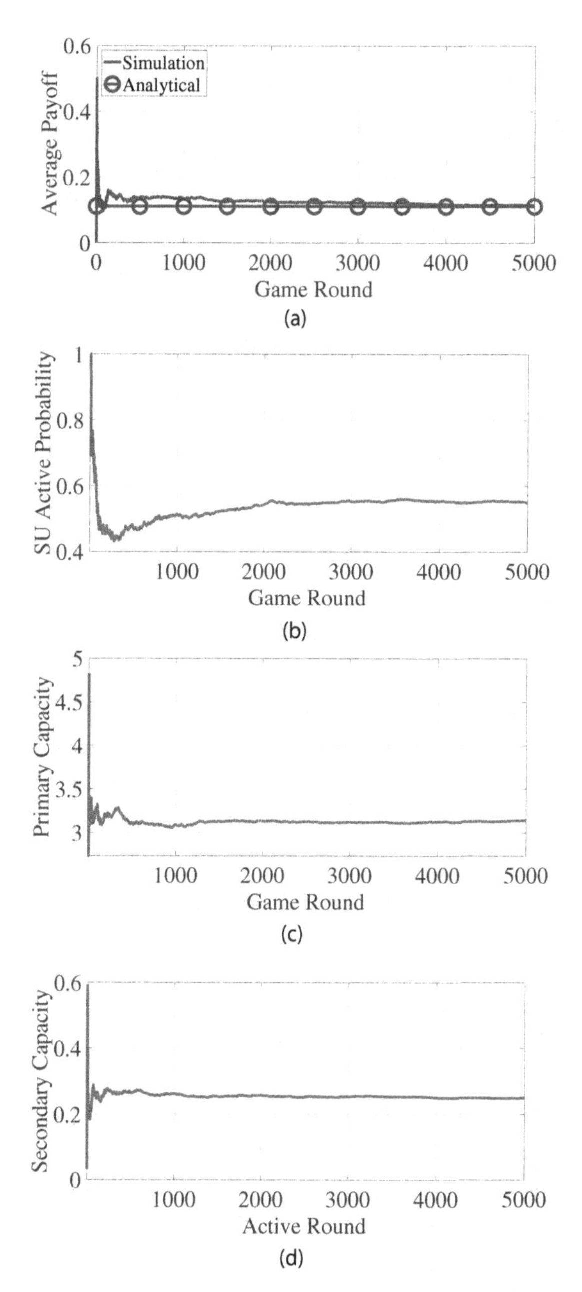

Figure 7.8 Long-term performance of the primary user showing different metrics including average outage probability and channel capacity. (a) Average payoff (i.e., outage probability of the primary user). (b) Probability of the IoT device of interest being *Active* in a game round. (c) Average channel capacity of the primary user. (d) Average channel capacity of the IoT device of interest.

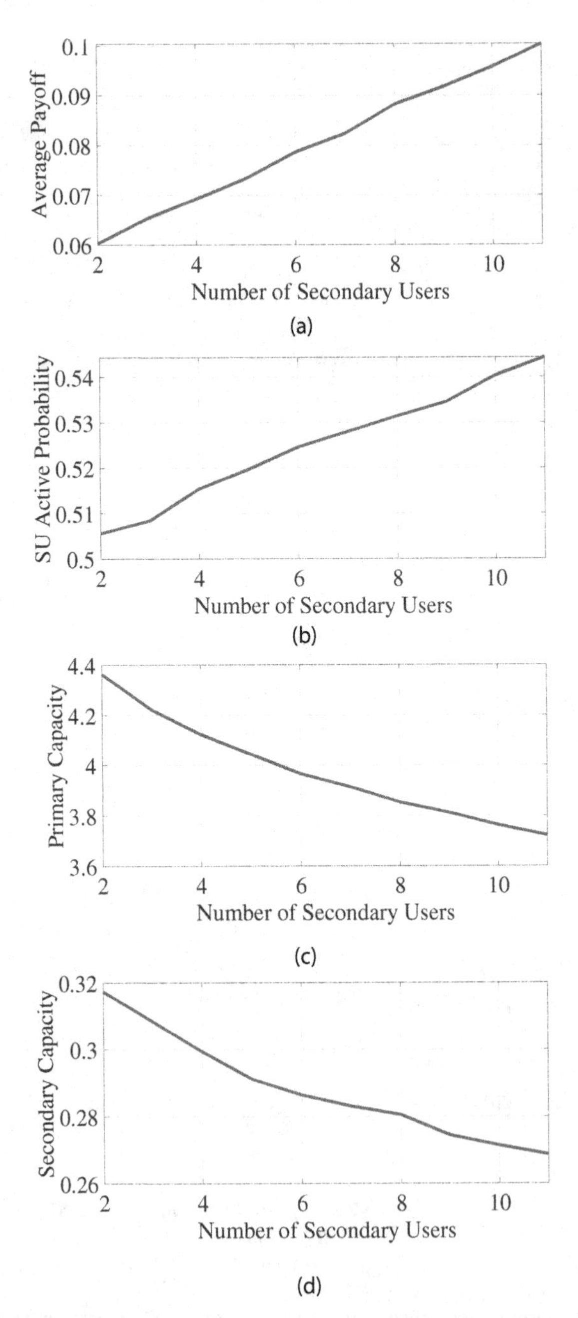

Figure 7.9 Impact of number of secondary users on the performance metrics of the algorithm. (a) Impact on the outage probability of the primary user. (b) Impact on the probability of the IoT device of interest being *Active* in a game round. (c) Impact on the channel capacity of the primary user. (d) Impact on the channel capacity of the IoT device of interest.

In the proposed framework, IoT devices do not need to know the full transmission history of the other devices in order to employ the transmission strategy and satisfy the QoS constraints. In addition, an IoT device perform transmission actions based on available channel-access information from the previous transmission interval. Hence, IoT devices benefit from an uncoordinated spectrum-sharing channel-access model while maintaining quality-of-service (QoS) constraints.

The proposed approach of using cognitive communications for IoT industrial WSN enables extending the number of secondary unlicensed users sharing resources with a primary user within allowed "soft" constrains. This approach to scaling can be related to CDMA scalability approach where increasing number of users softly deteriorates the system performance.

Numerical and simulation results illustrate the benefits of the proposed model in achieving data availability through uncoordinated IoT channel access. The proposed communication model appears suitable for settings like industrial wireless sensor networks with limited-resource IoT devices. Future extensions of this work will investigate how this model can be extended for sensor/actuator IoT devices requiring bi-directional communication and data availability requirements.

REFERENCES

1. Bilal Afzal, Sheeraz A. Alvi, Ghalib A. Shah, and Waqar Mahmood. Energy efficient context aware traffic scheduling for IoT applications. *Ad Hoc Networks*, 62:101–115, 2017.
2. Ashraf Al Daoud, George Kesidis, and Jorg Liebeherr. Zero-determinant strategies: A game-theoretic approach for sharing licensed spectrum bands. *IEEE Journal on Selected Areas in Communications*, 32(11):2297–2308, 2014.
3. Mohammed M. Alani and Ali Ismail Awad. An intelligent two-layer intrusion detection system for the internet of things. *IEEE Transactions on Industrial Informatics*, 19(1):683–692, 2022.
4. Bako Ali and Ali Ismail Awad. Cyber and physical security vulnerability assessment for IoT-based smart homes. *Sensors*, 18(3):817, 2018.
5. Ali Ismail Awad and Jemal Abawajy, editors. *Security and Privacy in the Internet of Things: Architectures, Techniques, and Applications*. John Wiley & Sons, Hoboken, NJ, 2021.
6. Waleed Ejaz and Mohamed Ibnkahla. Multiband spectrum sensing and resource allocation for IoT in cognitive 5G networks. *IEEE Internet of Things Journal*, 5(1):150–163, 2017.
7. Abdallah Farraj. Switched-diversity approach for cognitive scheduling. *Wireless Personal Communications*, 74(2):933–952, 2014.
8. Abdallah Farraj. Cooperative transmission strategy for industrial IoT against interference attacks. In *IEEE Texas Power and Energy Conference (TPEC)*, College Station, TX, USA, pp. 1–6, 2023.
9. Abdallah Farraj. Coordinated security measures for industrial IoT against eavesdropping. In *IEEE Texas Power and Energy Conference (TPEC)*, College Station, TX, USA, pp. 1–5, 2023.

10. Abdallah Farraj and Eman Hammad. Impact of quality of service constraints on the performance of spectrum sharing cognitive users. *Wireless Personal Communications*, 69(2):673–688, 2013.
11. Abdallah Farraj and Eman Hammad. Performance of primary users in spectrum sharing cognitive radio environment. *Wireless Personal Communications*, 68(3):575–585, 2013.
12. Abdallah Farraj and Scott Miller. Scheduling in a spectrum-sharing cognitive environment under outage probability constraint. *Wireless Personal Communications*, 70(2):785–805, 2013.
13. Abdallah Farraj, Scott Miller, and Khalid Qaraqe. Queue performance measures for cognitive radios in spectrum sharing systems. In *IEEE International Workshop on Recent Advances in Cognitive Communications and Networking (RACCN): Global Telecommunications Conference (GLOBECOM) Workshop*, Houston, TX, USA, pp. 997–1001, December 2011.
14. Federal Communications Commission. Spectrum policy task force report. ET Docket No. 02-135, November 2002.
15. Andrea Goldsmith. *Wireless Communications*, first edition. Cambridge University Press, Cambridge, 2005.
16. Kamal Gulati, Raja Sarath Kumar Boddu, Dhiraj Kapila, Sunil L. Bangare, Neeraj Chandnani, and G. Saravanan. A review paper on wireless sensor network techniques in internet of things (IoT). *Materials Today: Proceedings*, 51(1):161–165, 2021.
17. Mardiana binti Mohamad Noor and Wan Haslina Hassan. Current research on internet of things (IoT) security: A survey. *Computer Networks*, 148:283–294, 2019.
18. Kayiram Kavitha and G. Suseendran. Priority based adaptive scheduling algorithm for IoT sensor systems. In *International Conference on Automation, Computational and Technology Management (ICACTM)*, London , UK, pp. 361–366, 2019.
19. Dong Li. Performance analysis of uplink cognitive cellular networks with opportunistic scheduling. *IEEE Communications Letters*, 14(9):827–829, 2010.
20. Ling Li, Shancang Li, and Shanshan Zhao. QoS-aware scheduling of services-oriented Internet of Things. *IEEE Transactions on Industrial Informatics*, 10(2):1497–1505, 2014.
21. Moustafa Mamdouh, Ali Ismail Awad, Ashraf A. M. Khalaf, and Hesham F. A. Hamed. Authentication and identity management of IoHT devices: Achievements, challenges, and future directions. *Computers & Security*, 111:102491, 2021.
22. Fatma Masmoudi, Zakaria Maamar, Mohamed Sellami, Ali Ismail Awad, and Vanilson Burégio. A guiding framework for vetting the internet of things. *Journal of Information Security and Applications*, 55:102644, 2020.
23. Joseph Mitola. Cognitive radio: An integrated agent architecture for software defined radio. PhD thesis, Royal Institute of Technology (KTH), Stockholm, Sweden, December 2000.
24. Amitav Mukherjee. Physical-layer security in the internet of things: Sensing and communication confidentiality under resource constraints. *Proceedings of the IEEE*, 103(10):1747–1761, 2015.
25. Nahla Nurelmadina, Mohammad Kamrul Hasan, Imran Memon, Rashid A. Saeed, Khairul Akram Zainol Ariffin, Elmustafa Sayed Ali, Rania A. Mokhtar, Shayla Islam, Eklas Hossain, Md Hassan, et al. A systematic review on cognitive radio in low power wide area network for industrial IoT applications. *Sustainability*, 13(1):338, 2021.

26. Stephen S. Oyewobi, Karim Djouani, and Anish Matthew Kurien. A review of industrial wireless communications, challenges, and solutions: A cognitive radio approach. *Transactions on Emerging Telecommunications Technologies*, 31(9):e4055, 2020.
27. Maria Rita Palattella, Nicola Accettura, Luigi Alfredo Grieco, Gennaro Boggia, Mischa Dohler, and Thomas Engel. On optimal scheduling in duty-cycled industrial IoT applications using IEEE802.15.4e TSCH. *IEEE Sensors Journal*, 13(10):3655–3666, 2013.
28. Sanjeevi Pandiyan, Samraj Lawrence, V. Sathiyamoorthi, Manikandan Ramasamy, Qian Xia, and Ya Guo. A performance-aware dynamic scheduling algorithm for cloud-based IoT applications. *Computer Communications*, 160:512–520, 2020.
29. William H. Press and Freeman J. Dyson. Iterated Prisoner's Dilemma contains strategies that dominate any evolutionary opponent. *Proceedings of the National Academy of Sciences*, 109(26):10409–10413, 2012.
30. Nouman Shamim, Muhammad Asim, Thar Baker, and Ali Ismail Awad. Efficient approach for anomaly detection in IoT using system calls. *Sensors*, 23(2):652, 2023.
31. Zhengguo Sheng, Shusen Yang, Yifan Yu, Athanasios V. Vasilakos, Julie A. McCann, and Kin K. Leung. A survey on the ietf protocol suite for the internet of things: Standards, challenges, and opportunities. *IEEE Wireless Communications*, 20(6):91–98, 2013.
32. Ali Hassan Sodhro, Ali Ismail Awad, Jaap van de Beek, and George Nikolakopoulos. Intelligent authentication of 5G healthcare devices: A survey. *Internet of Things*, 20:100610, 2022.
33. Statista Research Department. Internet of Things (IoT) connected devices installed base worldwide from 2015 to 2025, 2016.
34. Heejung Yu and Yousaf Bin Zikria. Cognitive radio networks for internet of things and wireless sensor networks. *Sensors*, 20(18):1–6, 2020.

8 Review on Variants of Restricted Boltzmann Machines and Autoencoders for Cyber-Physical Systems

Qazi Emad Ul Haq
Naif Arab University for Security Sciences (NAUSS)

Muhammad Imran
Federation University

Kashif Saleem
King Saud University

Tanveer Zia
Naif Arab University for Security Sciences (NAUSS)

Jalal Al Muhtadi
King Saud University

8.1 INTRODUCTION TO RBMs AND AUTOENCODING

Restricted Boltzmann machines, more commonly known as RBMs, are undirected graphical models that are used as an integral part of the deep learning framework. RBMs were initially introduced in 1986 by the name Harmonium by Sumit Misra [1]. One of the most recent uses of RBMs was in the Netflix Prize; it outnumbered the competition in collaboration and filtration of data. The RBMs algorithm is used to reduce dimensionality, regressed classification, collaborative filtering, feature learning, and top modeling. RBMs algorithms work by learning first and then providing a closed-form depiction of the distribution pinpointing the observation. It is also used as a comparison of unseen probabilities of the observation alongside a sampling of learned distributions. A good example of this would be when we can repair some units that are seen and that correspond to a partial observation and then proceed

 DOI: 10.1201/9781003199410-8

to sample the remaining units that are visible for completing the observation. Even though RBMs were introduced back in the 1980s, with the technological advances their use is more relevant now. RBMs gained a lot of popularity in recent years after they were proposed for being used as the foundation of multi-layer learning systems known as deep belief networks. The concept behind this is that invisible neurons extract the features most relevant to the observation. These features can further be used as inputs to other RBMs. When RBMs are stacked like this, they can lead to exponential learning of multiple features that can help in reaching higher levels of representation.

Now, this brings us to the question of what autoencoders are. The simple explanation would be that autoencoders are a simple three-layer neural network. In this network, the output units directly connect with the input units. The number of invisible units in autoencoders is lower than the visible units, and the task of learning/training is to reduce errors when reconstructing, which means figuring out the most efficient and concise representation of output data.

RBMs share a similar belief; however, it utilizes imaginary units with a particular distribution rather than predetermined distribution. The main thing is to figure out how these two variables are connected to each other. One of the main features that differentiate RBMs and autoencoders is that RBMs have two very distinct biases [2]:

1. The invisible bias assists the RBMs to create activation on the forward pass.
2. The second bias is that the visible layers' biases aid the RBM determine reconstructions on the backward pass.

The rest of the chapter is organized as follows: Section 8.2 presents the background; Section 8.3 describes the malware attack detection using cyber-physical system (CPS). Section 8.4 discusses fraud and anomaly detection. Section 8.5 presents breakthroughs in CPS and their findings, while Section 8.6 discusses the presence of CPS as critical in the modern world, and Section 8.7. presents the evolution of CPS and its associated impacts. Section 8.8 concludes the study.

8.2 BACKGROUND

8.2.1 TARGETED PROBLEMS USING RBM's AND AUTOENCODERS

Most recently RBM models are directed toward the recognition, prediction, and classification of information. In recent years RBMs have been used for face recognition and detection. This helps in getting better results as compared to the latest modern techniques. RBMs can create deep systems for learning features. In simple words, the visible layer of the RBM model is the images, and the invisible layer will be the information for useful features. The input layer transmits image data to the invisible layer, which forces the hidden layer to describe it back to the input layer. However, in this environment, the invisible or hidden layers are restricted, which means that nodes in the concerned layer are unable to communicate with each other; they can only communicate with the nodes of the input layer. The input layer utilizes some loss of function to explain how badly the hidden layer is doing and suggest improvements to each node accordingly. We believe that after sufficient training passes, each node in the hidden layer will be able to focus on specific image characteristics that contain information and explanation of the images. Another way to think of this is to imagine that one node could learn to focus and determine the color of the eyes, another node could learn to focus on the shape of the lips, and so on.

The RBM model takes each visible node while simultaneously taking a low-level feature from a dataset to be learned. Node 1 is the invisible layer where x is multiplied by a weight and included in the list of biases. The result that is generated from this calculation is used to utilize an activation function, which further produces the output for the node and/or how strong the signal passing through it will be, as long as the input is x. When we have multiple inputs, we would combine them at one single hidden node. The same concept of multiple nodes with x and a separate weight will apply. The products are added and included in the bias, resulting in an activation function and producing the node's output.

Another targeted problem solved by RBMs is classification in statistics. Classification is the problem of identifying what sub-categories a particular observation belongs to, for example, classifying emails into spam and non-spam categories or assigning a diagnosis for a patient based on specific characteristics (e.g., gender, age, blood type, and the presence or absence of particular symptoms). An RBM algorithm implements classification referred to as a classifier. The term classifier refers to an RBM algorithm that maps input data to an output category.

Autoencoders are also used to reduce dimensionality, which means autoencoders are used as a pre-processing step for dimensionality reduction for performing fast and accurate reduction in dimensions without losing too much information. Certain dimensionality reduction methods can only perform linear dimensionality reduction, whereas complete autoencoders can perform larger-scale more nonlinear reductions in dimensionality. Another problem solved with autoencoders is the de-noising of images; they achieve this by not distinctly searching for the noise, unlike traditional methods; instead, they extract the image from noisy data which has been fed as an input through a learning representation. Once the input process is complete, the representation is decompressed to achieve a result of a noise-free image. Autoencoders are also commonly used to generate data for both image and time series. This is done by creating a distribution parameter at the bottleneck of the autoencoder which can be sampled at random to achieve discrete values to latent attributes that are then forwarded to the decoder, which results in the generation of image data. This can also be applied to model time series data like music.

TensorFlow [3], a deep learning library, and Scikit Learn, a machine learning library [4], both were implemented in the experiment discussed as follows: The experiment employed a seven-layer autoencoder with a 28-dimensional input that passed through a ten-neuron input layer that connected a ten-neuron layer to a five-neuron hidden layer. As an activation function, rectified linear units (RELUs) were utilized between the many layers of the autoencoder. A gradient descent technique was used to train the autoencoder.

Self-driving cars are another innovation that is powered by artificial intelligence. The designers of autonomous cars employ numerous data from image recognition systems along with machine learning techniques and neural networks to power autonomous driving cars. Neural networks identify trends in the dataset, which are used as input for machine learning algorithms. The input data include images from the inbuilt cameras on self-driving cars of other cars on the road, pedestrians, traffic lights, streetlights, and other important things the autonomous car needs to look out for on the road [5].

Over the recent years, we have seen an increase in the number of attacks taking place on CPS resulting in exponentially harmful consequences. Some of the common forms of cyber-attacks are eavesdropping, where secure information is interjected and a third party can listen to important and confidential information being transmitted, leading to a violation of security and in some cases even more serious consequences affecting numerous individuals [6].

Supervised learning and unsupervised learning are the two basic types of deep learning methodologies. The use of labeled training data is the difference between these two methodologies. Convolutional neural networks (CNNs) [7] that use labeled input belong within the category of supervised learning, which uses a particular architecture for image recognition. Deep belief network (DBN) [8], recurrent neural network (RNN) [9], autoencoder (AE) [10], and its derivatives are examples of unsupervised learning approaches. Following that, we go over some recent research that is similar to our work, most of which is based on the KDD Cup 99 or NSL-KDD datasets [11]. The NSL-KDD dataset has been used in studies on intrusion detection [8,12,13]. In a software-defined network (SDN) context, Hnat et al. [8] employed deep neural networks (DNNs) to create an anomaly detection model. They used six fundamental features from the NSL-KDD dataset's 41 features to train their model.

After dimensionality reduction, they utilized DBN to extract features for intrusion detection and SVM to categorize the data. When compared to employing SVM or DBN as independent classifiers, the researchers' hybrid DBN + SVM technique improves the detection performance. To detect zero-day attacks with high accuracy, it is suggested that two deep learning-based anomaly detection models employ AE and denoising AE. They also employed a stochastic strategy to establish the threshold value that has a direct impact on the suggested models' accuracy [13].

Another example is self-taught learning (STL), which is an effective and adaptable NIDS that combines a sparse AE for unsupervised dimensionality reduction with SoftMax regression to train the classifier. In two-class, five-class, and 23-class classification problems, our technique demonstrated satisfactory classification accuracy.

8.2.2 TECHNIQUES USED FOR CYBER-PHYSICAL SYSTEMS USING RBMs AND AUTOENCODERS

With technological advancements, businesses and government organizations are moving toward using online platforms for transactions and storing data on cloud servers, also known as e-commerce. These technological advancements have improved the productivity of banks, telecommunication companies, retail stores, health insurance, and other online businesses. However, along with improving productivity, it has also led to these businesses becoming more susceptible to hackers and other fraudsters. This has resulted in increased financial fraud and loopholes in cybersecurity.

Deep learning has empowered security in cyber-physical systems, by creating fraud detection models. These models work with biased data along with any insignificant features present in the input. These multiple attributes present themselves as an obstacle for the classifier to properly learn from multiple sources of data. Figure 8.1 shows an example of a basic four-input neural network.

To solve these problems, a dual-step method was proposed, and this method is split into two stages; in the first stage, a lesser dimension of features is taken from the input, and in the following stage, the model determines whether the transaction was a fraud or not. Simultaneously, a model for determining fraud is proposed based on autoencoders, where they extract important features from the input pool of data preceded by an algorithm used for classification. To detect credit card fraud, the method focuses on debit and credit card transactions. A transaction can have multiple sources that take into account the time and amount of the transaction, whether the transaction was a deposit or withdrawal, details of the customer, and/or not limited to the location of the ATM. These numerous attributes can result in the bad performance of algorithms. Since real transactions can have multiple features and high dimensions of data, it makes dealing with some data tedious. The crucial objective is to extract important data and remove the noise of extra data. To address these issues, the researchers are using autoencoders that can efficiently make a low dimension of the depiction of the input data while simultaneously having the ability to determine

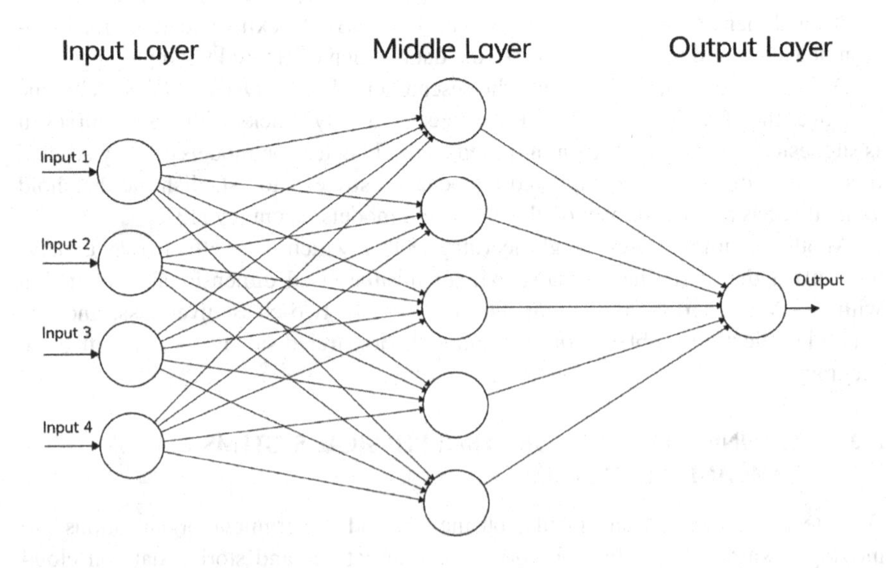

Figure 8.1 Basic 1 middle layered neural network architecture [14].

the nonlinear relation of features. Autoencoders depend on feed-forward neural networks that are further used for efficiently understanding encodings for training data.

The autoencoder network is equipped with the same source of input and output amplitude. It helps to understand the data that are used as input into an invisible depiction. With different amplitudes of the input and output data, it can reconstruct the input from the invisible representation. In simple words, it tries to identify the approximate function that has made the problem easier. However, by placing limitations on the network and limiting the number of invisible units, the irrelevant solution can be discarded. A general structure of the autoencoder-based model is shown in Figure 8.2.

An RBM is an artificial neural network technique, and it was at first designed only for autonomous learning activities that exploited raw data to determine adaptable patterns. RBMs are energy-based models that also employ a layer of invisible variables to model a distribution over visible variables [15]. Each variable can only take a binary value of 1 or 0 as represented a variational approach to unsupervised learning as shown in Figure 8.3.

RBMs are used as the foundation for deep Boltzmann networks. They rely on the idea that concealed neurons extract relevant attributes from the features. Further, these features are employed as input to another RBM. Hence, stacking RBMs results in attributes learning from other attributes and resulting in a high level of interpretation.

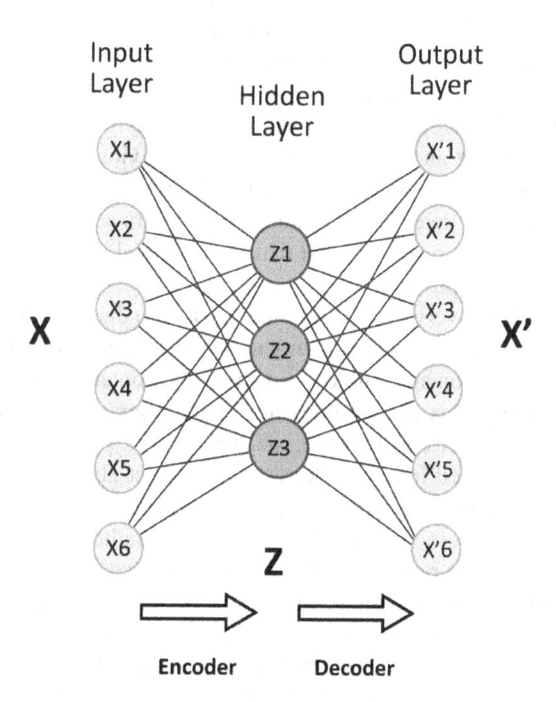

Figure 8.2 An autoencoder-based model for detecting fraudulent credit card transactions [1].

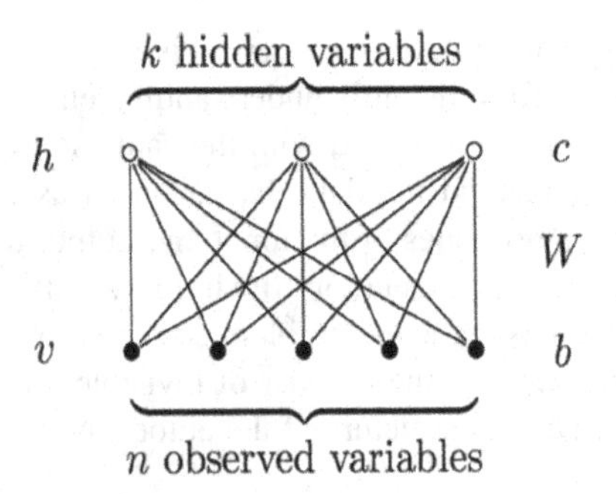

Figure 8.3 Variational approach to unsupervised learning with "k" hidden variables and "n" observed variables [11].

8.2.3 DETECTING NETWORK INTRUSIONS TO ENSURE THE SECURITY OF CPS IN IoT DEVICES

Cyber-physical security is one of the major problems that we are facing in the 21st century; one of the efforts to solve this problem utilizes intrusion detection systems along with firewalls. The consolidation of these two systems helps prevent the leakage of information from web applications. The IDS method is based on two groups, namely, detection of anomalies and signature detection. The signature approach is simple and works effectively when it comes to detecting attacks. However, it is not efficient in facing new attacks; on the contrary, anomaly detection approaches search for a pattern that does not conform to the usual pattern – this is established via machine learning and pattern recognition methods. These designs can easily distinguish new types of interference and detour from normal behavior. There is a set of steps that are followed for anomaly detection, which include feature development, attribute preference, and anomaly detection classifier. These characteristics increase in a direct proportion to n of the n-gram model making it critical to contemplate results to deal with dimensionality problems [16].

The Knowledge Discovery and Dissemination (KDD) dataset, 1999, is one of the most widely used datasets for detecting any network intrusion [17]. The dataset was generated for a competition and employs a dataset that contains over 4 million traffic records. Unfortunately, this dataset does not include data on raw network traffic. Based on the fundamental content type, and traffic type parameters, raw network traffic data are reprocessed into 41 features. The dataset also contains 22 different types of intrusions that may be divided into four categories: denial of service (DoS), unauthorized remote access from machines (R2L), unauthorized access for a user to root attack (U2R), and probe. Another examination reveals that the dataset has several flaws. The counterfeit nature of the network and attack data, as well as lost data due to advances and other imprecise definitions, are major contributors. There are also a lot of unnecessary entries, which causes the data to be skewed; as a result of these faults, a new dataset called NSL-KDD was proposed. This is also a common technique for identifying network interferences.

A few datasets consist of raw packet data, and the most widely used is the CTU-13 dataset [18]. This dataset comprises raw pcap files that identify the malignant, background, and normal data. The advantage of using raw pcap files as compared to the KDD 1999 and NSL-KDD dataset provides parties the chance to pre-process, using a larger number of algorithms. In addition to this, the CTU-13 dataset is unsimulated. Botnet attacks take place, and the anonymous traffic is from a larger network with no truth related to it. It is a mixture of 13 various sketches with multiple systems and several contrasting botnets, which makes it a very complicated dataset. Domain generation algorithm (DGA) detection has three prominent sources of data. Previously, machine learning research studies have shown a diversity of feature derivation methodologies, and the DL algorithms are primarily dependent upon domain names. This reflection can be seen in the three primary sources of DGA.

8.3 MALWARE ATTACK DETECTION

Malware attacks are the most common and continually increasing, which only makes it more difficult to block them using conventional procedures. Deep learning offers a chance to make generic models to find and analyze viruses automatically. This can result in providing defense against minor malware using unknown viruses and large-scale actors by employing innovative types of viruses to attack organizations and persons. There are numerous ways to detect malware such as developing DL-based detectors of hostile Android applications that use features from steady and dynamic analysis. Three key sources drive these features: static analysis with required permissions, sensitive application program interfaces (APIs), and active behaviors. In contrast, the vital behavior features are extracted from dynamic analysis where information is collected from driodbox, which is an application sandbox for the operating system of Android. These traits are further used as input data to DBN with two invisible layers that achieve ~97% accuracy, ~98% TPR, and ~4.5% FPR [18]. Multiple settings have been put to test, and out of all, the most consistent has been a two-hidden-layer DBN.

This proves that variations are more dependable than static features, making API calls, which are derived from running software in a sandbox, commonplace. Pascanu et al. [19] constructed a viral detection system that combines RNNs in conjunction with a multilayer perceptron (MLP) with regression models for classification. RNN has been instructed without supervision to predict the next API call. The proceeds of the invisible RNN serve as an input to the classifier after carrying out max-pooling on the trait vector to avoid the potential reordering of temporal events. To make sure the features contain cellular patterns, the invisible state from the middle of the sequence and the final invisible state are used.

Kolosnjaji et al. [20], on the contrary, employ CNNs and RNNs to locate the virus. A one-hot encoding is used to convert a sequence of calls to an API kernel into binary vectors. One-hot encoding is a method for gathering and classifying data in a way that is easy for machines to understand. The information gathered by this approach is then utilized to train deep learning algorithms that include CNN and RNN, as well as LSTM or SoftMax layer. Others have also employed API calls to create a deep-learning virus detector [21], which employed autoencoders with a sigmoid grouping layer for the task and attained ~96% accuracy.

8.4 FRAUD AND ANOMALY DETECTION

Credit cards were originally introduced to the banking industry a little more than 10 years ago. Credit cards are now a common method of payment for online purchases of goods and services. The users used these credit cards to make personal payments and other online transactions. As a result, the bulk of malware attacks has concentrated on template matching, which finds unusual patterns as opposed to regular transactions. Many methods for identifying fraudulent activity have been introduced in recent years. We will discuss these methods in detail in the coming paragraphs [13]. Credit card fraud can be detected using various methods ranging from generic machine learning models to state-of-the-art deep learning-based methodologies. One such method is the use of K-nearest neighbor (KNN) algorithms. This is a technique for supervised learning. By computing its nearest point, KNN is used to classify credit card fraud detection. If a new transaction is carried out, and the point is close to a fraudulent transaction, KNN classifies it as fraud [22]. The terms K-means and KNN are not interchangeable, and K-means clustering is an unsupervised learning approach. By clustering the data into groups, K-means aims to find new patterns in the data. In contrast, KNN is the number used to classify or forecast a new transaction based on the previous history by comparing the nearest neighbor.

In KNN, the distance between two data instances can be determined using a variety of methods, the most common of which is the Euclidean distance. KNN is a great resource. Defining the threshold K is the most difficult component of this process. We can use a high percentile of the data distribution if our data collection contains no anomalous cases. We can develop a validation set and optimize the value of threshold K on it if we have at least some anomaly examples in our data collection [23]. Overall, the network training and threshold K definition is a reasonably simple and quick operation, with a proper classification rate of 84% for fraudulent transactions and 86% for "regular" transactions. However, more effort could be done to increase performance, as is customary. The network parameters, for example, might be improved by experimenting with other activation functions and regularization parameters, increasing the number of hidden layers and units per hidden layer [23].

Outlier detection is another way of unsupervised learning by which malicious actors can be detected. The controlled outlier identification approach analyzes the outlier using the training dataset. The unmonitored biometric system, on either hand, is similar to categorizing data into many groups based on their features. In Ref. [22], the authors mentioned that unsupervised-learning outlier detection is preferable as compared to supervised-learning outlier detection for identifying credit card fraud since outlier detection does not require pre-existing information to label data as fraudulent. As a result, using example transactions, it is trained to classify between legitimate and illegitimate transactions [22] as shown in Figure 8.4.

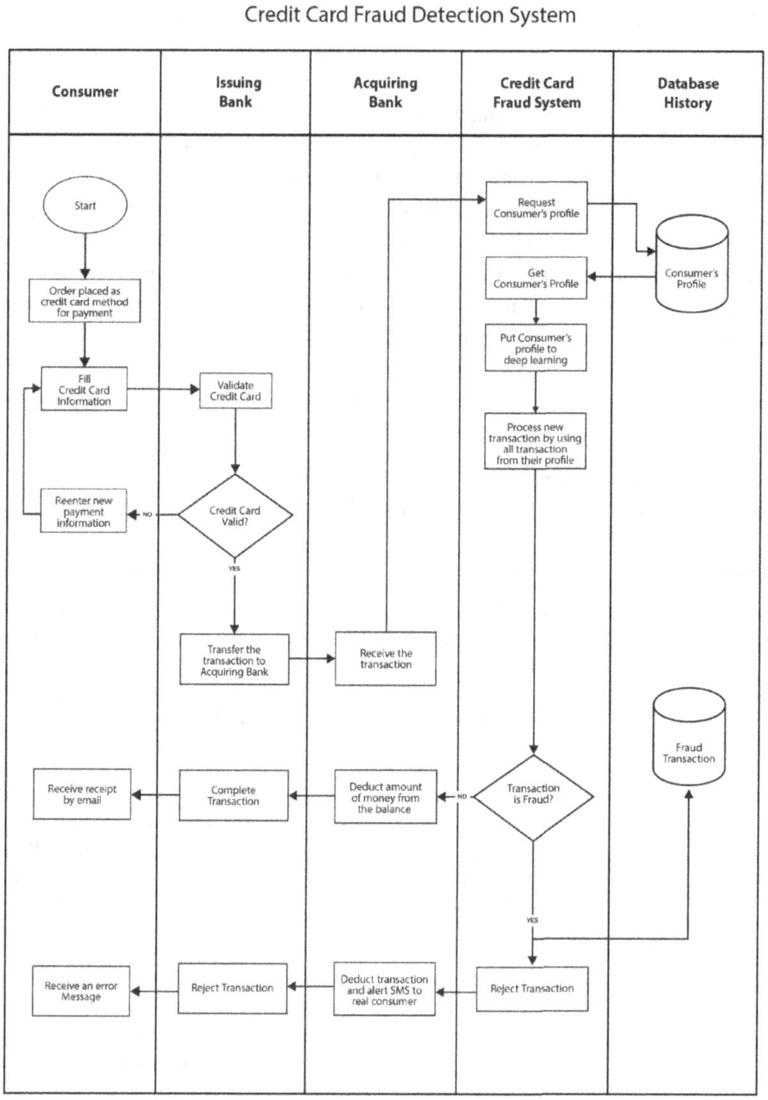

Figure 8.4 Analysis of credit card fraud identification techniques based on KNN and outlier detection [22].

One of the most advanced and evolving approaches is computer vision, which is a cutting-edge approach that has grasped the IT community's interest recently. Deep learning revolves around several hidden layers that increase the complexity of the model beyond human comprehension. In contrast, feed-forward computational models feature only one hidden state. Deep learning refers to multilayer neural networks. Many deep learning approaches, including AE, deep convolutional networks, support vector machines, and others, can be used. The main challenge with selecting an algorithm is that one must first comprehend the underlying problem and the goals of each deep learning approach for a specifically targeted problem.

There are no consistent patterns in fraudulent activities. They constantly alter their behavior, necessitating the use of unsupervised learning. Fraudsters gain access to new technology that enables them to commit scams via internet transactions. Consumers' habitual behavior is assumed by fraudsters, and fraud trends very quickly. Because some mischievous outfits conduct frauds once using online channels and then transition to other ways, fraud detection systems must detect online transactions using unsupervised learning [13], and a comprehensive comparison of various approaches used for fraud detection in CPS is presented in Table 8.1.

Table 8.1
Deep Learning-Based Methodologies on Autoencoder and Restricted Boltzmann Machine for Fraud Detection in Cyber-Physical Systems

Methodologies	Advantages	Limitations
K-nearest neighbors algorithm	The KNN method is simple to develop and can be used to detect anomalies in the target instance	With memory constraints, the KNN approach is ideal for detecting fraud
Hidden Markov Chains (HMM)	At the time of the transaction, HMM can detect fraudulent activities	With only a few transactions, HMM is unable to detect fraud
Neural network	Neural networks can detect real-time credit card scams since they have learned previous behavior	There are numerous sub-techniques in neural networks. As a result, if they discover something that isn't ideal for detecting credit card fraud, the method's performance will suffer
Decision tree	Nonlinear credit card transactions can also be handled by decision tree	Decision trees can be built using different induction algorithms such as ID3, C4.5, and CART, and they can have a variety of input features. As a result, the disadvantages include how to set up an induction method to detect fraud. DT is unable to detect fraud in the middle of a transaction
Outlier detection methodology	Outlier detection uses less memory and processing to detect credit card fraud	For huge online databases, this strategy is quick and effective. Other methods, such as outlier detection, are incapable of accurately detecting anomalies
Deep learning	Deep learning has several advantages, one of which is the ability to analyze and learn from large amounts of unsupervised data. It can extract complex patterns	Deep learning is now widely employed in the field of image recognition. There is no information available to explain the other domains. The deep learning library does not include all algorithms

Even in the absence (or with very few examples) of fraudulent transactions, the neural autoencoder provides a wonderful chance to create a fraud detector. The concept comes from the subject of anomaly detection in general, but it also works well for fraud detection [23].

Only regular data, in our case, and legitimate transactions are used to train a neural autoencoder with a more or less complicated architecture to reproduce the input vector onto the output layer. As a result, the autoencoder will learn to reproduce "normal" data properly [13].

The autoencoder's reconstruction of the input vector can be failed due to the anomalies. As a result, if we calculate the distance between the original data vector and the reconstructed data vector (the authors used a mean square distance), anomalies will have a considerably bigger distance value as compared to normal data. This is based on the distance between the input vector and the reconstructed output vector, and we can identify possibilities for fraudulent transactions. In terms of a rule, if the distance value exceeds a certain threshold K, we have a fraud/anomaly candidate [23].

By incorporating a bias in the definition of threshold K, the entire process might be driven to lean more toward fraud. In some cases, it may be preferable to accept a higher number of false-positive check-ups than to miss even one fraudulent activity [22].

8.5 BREAKTHROUGHS IN CPS AND THEIR FINDINGS

Scientists define cyber-physical security in different ways, and these definitions are usually based on their scientific perceptions. As we have already understood in the earlier text that CPS integration is based on numerous fields of science and engineering, cyber human systems (CHS) can be used to understand information at every stage as humans we have an inherent intelligence. This intelligence is easily misused for self-adaption, remedial, and precautionary actions. Integrating the element of nature and humanity is not an easy thing to recreate. CPS consists of two basic components; the first is a connection that guarantees real-time data collected that is gained from the physical world and then transmitted back as feedback received from cyberspace. The second component is managing intelligent data, analyzing, and computing capabilities that are used to construct cyberspace. To develop and deploy CPS for manufacturing, the 5C structure is used as a benchmark [23]. The 5C structure is explained in Figure 8.5.

The first and foundational step for creating an application for a cyber-physical system is to gather correct and reliable data from an engine and its parts, known as smart connections. The next step is to extract useful results from the information, for which there are several tools and concepts present for data conversion to information; this step is described as data-to-information conversion. The information collected and extracted in the first two steps then has to be accumulated as a center of information for this design through a network of machines. Then, we apply CPS using the

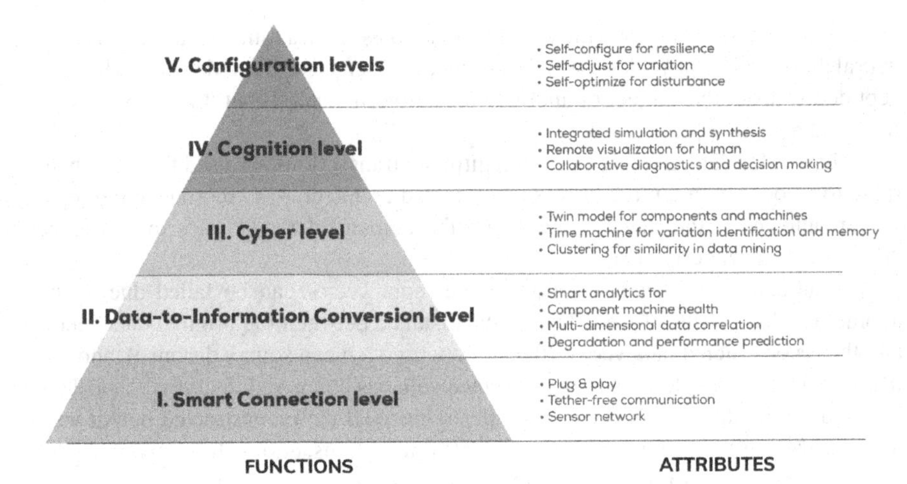

Figure 8.5 The 5C structure depicting the five-level hierarchy of employed during the deployment of a CPS [23].

results in a knowledge system that is monitored, and lastly, configuration assessment from cyberspace is implemented to tangible space. This physical space behaves as a managerial control for the self-configuration and self-adjustment of machines.

8.5.1 AIM OF A CPS-BASED SYSTEM

A CPS-based system aims to deploy the integration of cyber-physical and cyber humans – making the role of humans critical in the industry. Currently, the said technology is used for medical treatment, electricity power supply, smart homes, transportation, and agriculture. The power system used typical information fusion systems that can be characterized by bringing together physical systems and information systems.

The CPS structure is aimed at completely understanding and reflecting information and physical processes [8]. In the agriculture sector, a recent method of agricultural development is known as precision agriculture, which is a smart system that is supported by information technologies [8]. Precision agriculture is mainly impacted by factors that are not easy to predict as it is almost unattainable to create a system of linear differential equations. CPS has evolved and becomes an integrated system that provides discrete solutions through a hybrid system that can gather information from the external world and relate real-time and intelligent control to the real world, thus making the application of information a great success in the field of agriculture.

8.5.2 BREAKTHROUGHS IN CPS-BASED SYSTEMS

Other breakthroughs of CPS are seen in the field of medicine as the living standards of individuals become better because they start taking health more seriously.

Increased attention is being paid to research based on solving serious and new health issues. Medical CPS is an intelligent and network-based high technology-based system to ensure the safety of patients. This brings us to the point where knowledge needs to be controlled along with computers and medicines, for example, demonstrating a CPS application used to detect the state of a patient's breathing and breathing machine. A study was conducted on a medical device, which was a plug-and-play system used in laser tracheotomy. The field of medicine has evolved to become a field of increased research directed at using CPS [9].

Even though there are multiple ways and tools available which can be used for detecting threats, however, many of these tools have different skills resulting from the lack of a uniform approach and analysis, and because of these differences in the CPS, the generic modeling tools were not applicable [10]. Furthermore, maintaining security in CPS decision control systems including home security systems is a challenge that needs to be resolved [24–28]. We have already concluded that the CPS has an assortment of characteristics. The understanding of how communication of a CPS network system also has numerous network technologies such as Wi-Fi, internet, and wireless sensor networks. To achieve real-time and reliable interconnections and cooperation among different networks, there is still room for development and problems to be solved.

The testability of CPS is missing from the widely accepted validation standard. It is difficult for a CPS design to assess and confirm each component. It is also difficult to research a heterogeneous model's authentication mechanism and put it into practice for sophisticated heterogeneous systems. The macro lab has been chosen as CPS's core programming and verification environment, with promising results [8].

Because of CPS's complex and diversified properties, the system's security requirements have been raised. Individuals' livelihoods are dependent on system security, particularly in medical treatments, electricity, transportation, and other areas. Physical dangers and other threats posed by the area of information are the main sources of concern for CPS. The physical domain is primarily for the physical threat of assault, with the local node domain information and communication network attack being the primary threat. The current technology is insufficient to ensure good CPS security.

8.6 ENSURING CPS IS CRITICAL IN THE MODERN WORLD

CPS is still a relatively recent research area in the global information economy, and it has a significant impact on existing structures. Developing CPS is still in its early stages, regardless of whether you are based at home or overseas, and many challenges must be overcome. The future holds enormous promise in terms of CPS technology research and development, with an emphasis on system creation and modeling, verification, heterogeneous network difficulties, mass heterogeneous data processing and security, and much more.

Although 20th-century scientific advancements provided us with good approaches and instruments for building computational and physical systems. Designing a cyber-physical system is more than just combining two disciplines. Knowledge professionals have always had a vague awareness of the demands placed on them by the physical world. Mechanical, civil, and chemical engineers, for example, have viewed computers as instruments for solving algorithms while disregarding the physical properties of embedded computing systems. This has progressed to the point that it is now being used in the construction of cyber-physical systems. We will be able to construct new machines with complicated dynamics and excellent dependability due to a new discipline called cyber-physical system design. It will open the way for the reliable and cost-effective application of cyber-physical system principles to new sectors and applications.

However, over the last decade or two, we have accumulated a substantial body of evidence that supports the separation of information and physical sciences. For example, in most programming languages, the prevailing abstractions avoid explicitly representing time and other physical values, instead of grouping relevant physical design factors under the category of non-functional needs, which includes timing, power, and dependability.

On the contrary, engineering is increasingly relying on computer-based implementation, and systems have developed and evolved abstractions that disregard the quiet aspects of computation and communication. This inevitably acts as a barrier around systems and computer science, preventing the communities from combining their knowledge and, as a result, partitioning education into isolated disciplines, giving rise to segmented design flows that have resulted in substantial problems and failures as complexity grows. The current industrial experience demonstrates that our understanding of how one should collect computers and physical systems has reached its limits. We must continue to construct systems with our restricted methods and tools to adequately address issues for people and establish predictable systems.

8.7 EVOLUTION OF CPS AND ITS ASSOCIATED IMPACTS

Cyber-physical systems cover a wide range of application areas; we can use new scientific technology and technological understanding of interactions of processing information, networks, and physical processes. The new science of CPS will have a wide range of applications that will assist specializations and additions to specific application domains. This innovative technology of CPS will permit users to design systems more economically by using both abstract knowledge and advanced tools. It could also lead to the development of more reliable CPSs, which will allow us to apply benchmarked practices to a whole range of cyber-physical applications [12].

The use of CPSs in the real world is something that is of great importance to us in the 21st century, and CPSs can be utilized in controlling industrial systems. These control systems can optimize the production process while simultaneously overseeing other systems such as sewage systems or nuclear plants or irrigation plants used for agriculture purposes. The CPS controller apparatus can have multiple functionalities that will allow it to work in coherence and result in better more efficient reporting of outcomes. Currently, different connections are being used to bridge the gap, between these apparatuses, and between the virtual and real world. If there was ever a situation where the power grids were hacked and security was compromised, the end user could lose power and power-providing companies would bear a financial loss; to avoid a situation like this, incorporating CPSs for virtual security has become a more crucial need. Most CPSs are not designed while prioritizing security as they are usually not interlinked with the internet, which ultimately makes tangible security the easiest way to keep customers safe; however, as technology advances, we are observing an increasing demand to incorporate virtual security to keep customers data safe from any type of criminal activity. As the use of the internet becomes more common, ensuring the safety of users will become more difficult and require state-of-the-art algorithms to accomplish this [13].

In the near future, we can see CPS will be beneficial for preventing chaos, which is ensured due to natural calamities; multiple technological resources support swift evacuation; an example of such systems would have been developed in the COVID-19 pandemic when the world was affected in a dire way; a CPS could have helped gauge and monitor the spread of the virus and created hot spots to alert people from avoiding certain areas where the virus was spreading faster than other areas. This could have resulted in better controlling the situation and controlling the number of impacted individuals.

Another way CPS is used today is in AI-powered dash cameras that are commonly used in the trucking industry across the United States. These cameras help predict and prevent incidents before they happen, by providing managers in a trucking company with data regarding the behavioral trends of truck drivers. The algorithms learn what behaviors are safe and what behaviors are considered risk factors or unsafe and accordingly record video footage and upload it to the cloud notifying managers and allowing them to improve the safety of their fleets by using coaching and mitigating the chances of future collisions. The CPS can also be used to help transport companies manage their fuel costs, the system has been designed in a way to calculate the optimal fuel usage for a particular vehicle type and generate reports to provide insights into whether the vehicle in consideration is utilizing fuel efficiently or not – minimizing the cost of companies and ultimately maximizing profits [13].

With the fast-paced economy, the attributes of CPS can come in handy to businesses across the board as it provides us with an opportunity to receive information in real time and minimize the time it takes to decide. Humans today are hungry for information; we want to know everything that is going on around us and this information can only be accessed through solutions offered by the CPSs. Today, it is much harder for a criminal to get away undetected; with our smartphones and social media, monitoring our every move and storing data to analyze and make customer experiences better, it is almost impossible to go undetected. The concept of big data also revolves around the relationship between the physical world and the non-physical world where our search histories determine our likes and dislikes and consequently present us with the options for food, travel, movies, etc., best suited to our personalities. We often wonder how when we are staying someplace and suddenly all the advertisements on social media are started to come related to that place and we ultimately assume to think that our smart devices are listening to our conversations or keep eye on us – however that is not the case. The bridge between the physical and cyber world has become so small that now when we connect to a Wi-Fi connection that belongs to a friend or a family member, we start seeing advertisements for the items or brands they've searched. This proves how CPSs can be used to fuel marketing campaigns and help companies reach a wider audience. Businesses that are investing in the CPS are gaining benefits with a high market share [29].

Although most of the uses of the CPSs have positive results, unfortunately, they can also be used for destruction and war. Wars are being fought using drones in today's day and age – these drones are controlled using cyber-physical systems, they can identify and disengage weapons from a distance without being noticed, and they are used to attack threats and even high-profile individuals. The use of such drones has ultimately led to anti-drone systems which have the main responsibility of preventing drone attacks, basically fighting machine against machine. This also raises the concern of infiltration into sovereign states and waging war without being detected – making the need for both physical and cybersecurity even more imperative. In the old days, armies were responsible for the security of a state, today we have surveillance cameras and drones that monitor any shady activity while also being prepared to attack or respond to an attack. Drone used for spy purposes also poses a serious threat to national security as these drones can go undetected by regular radars, which means countries not investing in updating their systems are prone to a technological threat at any given time [30].

Air transport will also see innovation at a large scale, even though currently air transport is considered one of the safest means of travel, but with CPS integrations air transport has room to develop and become even safer. With advancements in technology, we are looking at satellite monitoring real time of all airborne bodies. All aircraft and airports will be connected through a next-generation smart system which will make air travel less time-consuming, more efficient, and less harmful to the environment. CPS will encourage more automation in terms of air control and communication between the personnel on the ground and airborne. We are looking at balancing the roles of humans and machines, with a CPS, a lot of intricate aspects of air control that could be overlooked by human error will be monitored with more detail, making air travel and communication safer and more reliable. These advanced systems will also make maintaining the maintenance of aircraft more streamlined as we will be utilizing aspects of the human mind and machine capabilities in congruence eliminating the risk of human errors and accidents [31].

8.8 CONCLUSION

The plan of developing a more reliable CPS is the need of the hour. Development of a robust CPS requires continuous investment and collaborations to build systems that are domain and application-specific to solve a particular problem that may include fraud detection, network anomaly, and network intrusion detection to name a few. Ensuring IoT security through the employment of advanced CPSs will provide enhanced security to venerable systems. Cyber-physical systems are heterogeneous mixtures. They are a synthesis of computing, communication, and physical dynamics. They are more difficult to model, develop, and analyze than homogeneous systems. Because software interacts with physical processes in cyber-physical systems, the sequence of events in software can be significant.

Therefore, having a robust CPS that is based on state-of-the-art algorithms is extremely important. Improvements need to be made in CPS currently present to ensure an enhanced level of access protection to sensitive and vulnerable networks.

Investments in understanding and leveraging CPS are required to build advanced and reliable systems to monitor security and ease life by moving away from traditional methods and tools. Moreover, governments need to actively invest in a cyber-physical infrastructure that will operate more effectively and reliably. This will help serve the issue of both CPS technologies and enable increased use of cyber-physical techniques in other parts of the economy.

ACKNOWLEDGMENT

The author would like to express their deep thanks to the Vice Presidency for Scientific Research at Naif Arab University for Security Sciences for their kind encouragement of this work.

REFERENCES

1. Misra, S., Thakur, S., Ghosh, M., and Saha, S. (2022). An autoencoder based model for detecting fraudulent credit card transaction.
2. Upcommons.upc.edu (2022). [online] Available at: https://upcommons.upc.edu/bitstream/handle/2117/355516/survey_DL_cyber+-+final.pdf?sequence=3 [Accessed 10 March 2022].
3. TensorFlow (2022). TensorFlow. [online] Available at: https://www.tensorflow.org/ [Accessed 10 March 2022].
4. Scikit-learn.org. (2022). scikit-learn: Machine learning in Python: scikit-learn 0.16.1 documentation. [online] Available at: http://scikit-learn.org/ [Accessed 11 March 2022].
5. TechTarget (2022). What are self-driving cars and how do they work? Retrieved 11 March 2022, from https://www.techtarget.com/searchenterpriseai/definition/driverless-car.
6. Yaacoub, J.-P. A., Salman, O., Noura, H. N., Kaaniche, N., Chehab, A., and Malli, M. (2020). Cyber-physical systems security: Limitations, issues and future trends. *Microprocessors and Microsystems*, 77, 103201. [Ebook]. Retrieved from https://www.sciencedirect.com/science/article/pii/S0141933120303689.
7. Hussain, M., Qazi, E. -U. -H., AboAlSamh, H. A. and Ullah, I. (2023). Emotion Recognition System Based on Two-Level Ensemble of Deep-Convolutional Neural Network Models, *IEEE Access*, 11, 16875–16895, doi: 10.1109/ACCESS.2023.3245830.
8. Hnat, T. W., Sookoor, T. I., Hooimeijer, P., Weimer, W., and Whitehouse, K. (2008). Macrolab: A vector-based macro programming framework for cyberphysical systems. *ACM Conference on Embedded Network Sensor Systems*, pp. 225–238, Raleigh NC, November 5–7, 2008
9. Li, M. (2011). The development of foreign electronic medical records and its enlightenment to China. *Medical Information*, 24(6),1478–1481.
10. Xiang, W., Yang, C., and Chen, Y. (2015). A survey of modeling and simulation in CPS. *Journal of Chongqing Electric Power College*, 20(5), 43–47.

11. Shah, S. N. (2019). Variational approach to unsupervised learning. *Journal of Physics Communications*, 3. DOI: 10.1088/2399-6528/ab3029.
12. CPS Steering Group (2022). Retrieved 12 March 2022, from http://iccps.acm.org/2011/_doc/CPS-Executive-Summary.pdf.
13. Tyagi, A. K. and Sreenath, N. (2021). Cyber Physical Systems: Analyses, challenges and possible solutions. *Internet of Things and Cyber-Physical Systems*, 1, 22–33. [Ebook]. Retrieved from https://www.sciencedirect.com/science/article/pii/S2667345221000055.
14. Sanjab, A., Saad, W., and Basar, T. (2020). A game of drones: Cyber-physical security of time-critical UAV applications with cumulative prospect theory perceptions and valuations. *IEEE Transactions on Communications*, 68(11), 6990–7006. DOI: 10.1109/tcomm.2020.3010289.
15. Misra, S., Thakur, S., Ghosh, M., and Saha, S. (2022). An autoencoder based model for detecting fraudulent credit card transaction.
16. Vartouni, A. M., Kashi, S. S., and Teshnehlab, M. (2018). An anomaly detection method to detect web attacks using stacked auto-encoder. *In 2018 6th Iranian Joint Congress on Fuzzy and Intelligent Systems (CFIS)*, IEEE, pp. 131–134, Kerman, 2018
17. KDD Cup 1999 Data. (2022). Retrieved 11 March 2022, from http://kdd.ics.uci.edu/databases/kddcup99/kddcup99.html.
18. Stratosphere Lab (2022). Retrieved 11 March 2022, from https://stratosphereips.org/category/dataset.
19. Pascanu, R., Stokes, J. W., Sanossian, H., Marinescu, and M., Thomas, A. (2015). Malware classification with recurrent networks. In *Proceedings of the 2015 IEEE International Conference Acoustics, Speech and Signal Process, (ICASSP)*, Brisbane, Australia, 19–24 April 2015, pp. 1916–1920.
20. Kolosnjaji, B., Zarras, A., Webster, G., and Eckert, C. (2016). Deep learning for classification of malware system call sequences. In *Proceedings of the Australasian Joint Conference on Artificial Intelligence*, Hobart, Australia, 5–8 December 2016, pp. 137–149.
21. Hardy, W., Chen, L., Hou, S., Ye, Y., and Li, X. (2016). DL4MD: A deep learning framework for intelligent malware detection. In *Proceedings of the International Conference Data Mining (ICDM)*, Barcelona, Spain, 12–15 December 2016, p. 61.
22. Malini, N. and Pushpa, M. (2017). Analysis on credit card fraud identification techniques based on KNN and outlier detection. *In 3rd International Conference on Advances in Electrical, Electronics, Information, Communication and Bio-Informatics (AEEEICB17)*, Chennai, India.
23. Fraud Detection Using a Neural Autoencoder. (2022). Retrieved 12 March 2022, from https://medium.com/low-code-for-advanced-data-science/fraud-detection-using-a-neural-autoencoder-a5bdc244f390#:~:text=The%20idea%20stems%20from%20the, our%20case%2C%20 only%20legitimate%20transactions
24. ShieldSquare Captcha. (2022). Retrieved 12 March 2022, from https://iopscience.iop.org/article/10.1088/1757-899X/830/4/042090/pdf.
25. Awad, A. I., Furnell, S., Paprzycki, M., and Sharma, S. K., (eds.) (2021). *Security in Cyber-Physical Systems: Foundations and Applications*. Studies in Systems, Decision and Control. DOI: 10.1007/978-3-030-67361-1.
26. Hassan, A. M. and Awad, A. I. (2018). Urban transition in the era of the internet of things: Social implications and privacy challenges. *IEEE Access*, 6, 36428–36440. DOI: 10.1109/ACCESS.2018.2838339.
27. Ali, B. and Awad, A. (2018). Cyber and physical security vulnerability assessment for IoT-based smart homes. *Sensors*, 18(3), 817. DOI: 10.3390/s18030817.

28. Elrawy, M., Awad, A., and Hamed, H. (2018). Intrusion detection systems for IoT-based smart environments: A survey. *Journal of Cloud Computing*, 7, 21. DOI: 10.1186/s13677-018-0123-6.
29. Awad, A. I. and Abawajy, J. H. (2022). *Security and Privacy in the Internet of Things: Architectures, Techniques, and Applications.* Hoboken, NJ: Wiley-IEEE Press.
30. Pumsirirat, A. and Yan, L. (2018). Credit card fraud detection using deep learning based on auto-encoder and restricted boltzmann machine. *(IJACSA) International Journal of Advanced Computer Science and Applications*, 9(1), 1–8, 2018. [Ebook]. Retrieved from https://pdfs.semanticscholar.org/01be/7624aa0e0251182593350a98441 1c2e5128a.pdf?_ga=2.100901704.838168785.1649802794-2017936328.1649802794.
31. Ebook. (2022). Retrieved from https://www.nitrd.gov/pubs/CPS-OSTP-Response-Winning-The-Future.pdf.

9 Privacy-Preserving Analytics of IoT Data Using Generative Models

Magd Shareah and Rami Malkawi
Yarmouk University

Ahmed Aleroud and Zain Halloush
Augusta University

9.1 INTRODUCTION

The Internet of Things (IoT) is one of the essential areas that refer to the new multi-domain technologies to transmit data in real time. It consists of a global network of millions of physical devices in different places connected to the internet, collecting and sharing data, while having the ability to interact with each other [1]. This environment consists of devices and services and provides advanced levels of services. IoT systems now have a significant impact on how people manage many of their daily tasks in a more effective and efficient manner [2]. The scope of IoT applications is diverse, encompassing several aspects of human lives, such as smart cities, buildings, home automation, and wearable devices [2]. Integrating IoT technology in these different aspects has transformed the conventional ways of object functionality, which means more monitoring, communication, information sharing, and decision coordination with other nodes on the network. Recent studies show that the total number of interconnected devices in the IoT devices and units reached ~24 billion devices in 2020 [3].

Preserving data privacy in the IoT environment is a substantial challenge considering the high volume of transmitted and shared data between the different devices. The implementation of effective data-preserving methods requires high-end requirements in terms of storage, time, and energy consumption [4]. Moreover, most of the services offer vague terms and conditions that may ask users to agree to abandon their privacy and data protection and allow these service providers to use their metadata for commercial purposes, such as data analytics. Therefore, it is essential that users be fully aware of why, when, and for what reasons their data are being collected [5]. The IoT poses a significant challenge to the confidentiality of big data either generated, transmitted, or stored via IoT objects. To address this concern, privacy-preserving methods aim to maintain significant accuracy levels in analyzing such sensitive data using machine learning algorithms. Nevertheless, the volume and

 DOI: 10.1201/9781003199410-9

veracity of such data remain an obstacle to achieving this goal. One of the fields that is being used intensively in this area of research is privacy-preserving data mining for big data, which has proved that it can preserve data with satisfactory accuracy levels [6]. This chapter presents a novel approach for privacy-preserving data generation through generative models. This approach has gained popularity as an effective method of fitting generative models into real-life data generation problems. Recent studies show a promising trend for using these models due to their empirical performance. Those models are used to estimate the underlying distribution of a dataset and randomly generate realistic samples according to their estimated distribution [7].

Studies also show the potential of using such generative models in typical uses such as designing useful cryptographic primitives [8]. These models also proved superior in different fields and are widely used in biology and medicine to generate biomedical data [9]. However, using traditional generative models provides no guarantee on what the synthetic data reveal about participants. It is possible that the generator neural networks could learn to create synthetic data that reveal actual participant data. We will employ this type of generative model to create privacy-preserving techniques and produce data that could achieve privacy protection levels. In addition, these generative models provide a good advantage for analysts; it enables them to generate an unlimited amount of synthetic data for arbitrary analysis tasks without disclosing the privacy of training data. To achieve good performance, data mining methodologies require a significant amount of training data. However, collecting such data from specific domains (e.g., medical data) is often impractical due to privacy and sensitivity concerns. This makes building high-quality data analytics models a challenging task. Generative models provide a promising direction to alleviate the data scarcity issue. These models learn from the original data distribution and generate more samples for analysis studies. Generative adversarial networks (GANs) have demonstrated good performance in modeling the underlying data distribution [10]). This has been achieved by combining the complexity of deep neural networks and game theory to generate high-quality "fake" samples that are hard to be differentiated from real ones [11].

9.2 IoT ARCHITECTURE AND APPLICATIONS

IoT architecture consists of three layers; the first layer is the perception layer, also known as the sensor layer, which is implemented as the bottom layer in IoT architecture. This layer interacts with IoT physical devices and components through smart devices (e.g., RFID, sensors, and actuators). The main objectives of this layer are to connect nodes into an IoT network, to collect, process, and measure the state information associated with these nodes by the deployed smart devices. The processed information is then transmitted to the upper layer via a layer interface. The middle layer is the network layer, also known as the transmission layer. This layer receives the processed information transmitted by the previous layer, "i.e., perception layer," and determines the routes to transmit the data and information to the IoT hub, devices, and applications via integrated networks. This layer consists of integrated devices such as hubs, switches, cloud computing components, gateways, and various other communication technologies, such as Wi-Fi, long-term evolution (LTE), Bluetooth, and others.

Thus, it can be inferred that the network layer is the primary layer in the IoT architecture. Using interfaces and gateways, this layer transmits data from/to different applications among various networks and uses different communication technologies and protocols [12]. The third layer is the application layer, which represents the top layer in the IoT architecture. This layer provides services and operations according to the network layer's data; therefore, it is sometimes known as the business layer [13]. This layer encloses several applications such as smart grid, smart transportation, and smart cities, while each application encompasses different requirements [14].

The development of industry-oriented and user-specific applications is facilitated by IoT technologies. IoT applications need to ensure that devices receive data/messages and then act upon them properly on time. For example, FedEx uses Sense Aware to monitor the packages when it is opened and whether it was tampered with along the way and track the temperature, location, and other "vital" signs of the packages. Data visualization, on the contrary, is not necessary for device-to-device applications. However, in interactions with the environments, it is necessary to provide human-centered IoT applications to present information to end users in an intuitive and easily understandable way. IoT applications need to be built so that devices can monitor the environment, identify problems, communicate with each other, and potentially resolve problems without the need for human intervention [1].

9.3 LIMITATIONS AND CHALLENGES

IoT devices are uniquely identified. This usually creates security, privacy, and trust threats putting all connected devices at risk, thereafter, causing harm to the users of such devices. Since IoT devices collect personal information, such as user identity, location preferences, and activities, privacy threats are a major concern. The data collected can potentially disclose sensitive users' information and monitor their activities, even across other devices, thus compromising privacy through this data linkage process. To mitigate this problem, trust models should be built to avoid exposing data to public or private servers.

Privacy risks affect the adoption of IoT technology. For more control on data disclosure, data aggregation mechanisms and access policies represent prevention tools. Resource limitations are the main reason for the adoption of default privacy settings. However, there are three main technical limitations on the ability to secure the IoT environment. The first limitation is the heterogeneity of IoT systems, which complicates protocol design and system operations. The second is the scarcity of CPU and memory resources, which limits the use of resource-demanding crypto primitives, such as the public-key cryptography, which is used in most internet security standards. The third limitation is the end-to-end IoT-oriented security measures.

The aforementioned limitations can be listed under technical limitations in IoT architectures. There remain different types of privacy challenges that must be taken into consideration when addressing IoT structures. In this chapter, besides addressing some of the technical limitations, we are also focusing on several privacy challenges that are associated with the following threats:

- **User identification**: It is the ability to distinguish an entity or reveal his identity based on related data "identifiers and Quasi-identifiers" (e.g., name, address, or any other personal information). The wide adoption of IoT makes it easy to collect large amounts of data beyond users' control. Thus, multiple types of risks such as profiling and tracking individuals' behavior would appear. For example, many platforms' developers use machine learning techniques to infer personal information about users' interests and use this information to flood the user interface with targeted advertisements [15].
- **User tracking**: This risk primarily depends on the previously mentioned threat, "i.e., User Identification, " but this threat is based on location. This threat occurs when the data about a specific user are collected and then used to track users' behavior. After the user is identified, their location is also identified with their locations' history, which enables tracking. For example, the user who uses location-based services is required sometimes to share his location. So, sometimes tracking his location happens without his explicit consent and knowledge. Positioning techniques, which are based on global positioning systems (GPS), GSM RFID, and wireless LAN, also have a huge effect on tracking the users. This technology raises concerns about location privacy intrusion. Since personal information, including users' location, is collected by IoT devices and services, this information could be abused or sold to third parties for targeting advertisement purposes, and more seriously, some criminals may take advantage in exploiting such data and performing types of criminal activities that might compromise individuals' lives.
- **Profiling**: It is the process of recording and analyzing data to characterize personal behavior. This is used to assess or infer individuals' interest in a particular domain or group purposes. The rising number of internet-connected systems and the evolution of data mining algorithms and tools significantly contributed to the emergence of collecting users' big data. This was achieved using different data mining tools that establish a clear description of the customer needs and easily provide detailed customer profiles. Target advertisements, website personalization, and service matching are increasingly using profiling as a core advantage to enhance their business. Profiling can help learn and estimate the individuals' political and religious views and assess medical conditions [3]. This information about individuals can be shared without customers' consent, leading to privacy violation of the customers. So, limiting access to confidential information in the IoT environment will preserve the individual's privacy; however, it may negatively affect the accuracy of the data mining processes.
- **Utility monitoring and controlling**: This is related to the data collected about customers. Sensitive information can be used to infer users' life patterns, which cause privacy threats if acquired through unauthorized access, especially when

attackers get access privileges and use it to control usage in different IoT domains without the user's knowledge or permission. There are four main IoT domains, namely, home, enterprise, utilities, and mobile. At home applications, Wi-Fi is used, which provides high bandwidth and plays a significant role in video streaming services, and supports high sampling rates for audio streaming as well as control of appliances such as air conditioners, refrigerators, and washing machines [3].

9.4 IoT PRIVACY: DEFINITIONS AND TYPES

IoT aims to improve the overall quality of human lives by enabling a plethora of smart services in almost every aspect of our daily activities and interactions. This section provides a comprehensive overview of the most crucial privacy techniques that are widely used to preserve IoT data. Moreover, we also illustrate how different algorithms work, their applications, and the impact of each algorithm on both privacy and utility (data usefulness).

When discussing privacy, the concepts that illustrate it vary. In our context, we will use the privacy definition that was used in the Internet Security Glossary, which defines data privacy as "the right of an entity (normally a person), acting on its own behalf, to determine the degree to which it will interact with its environment, including the degree to which the entity is willing to share information about itself with others". From a broader perspective, privacy consists of seven main categories. These categories start with the privacy of the person, which includes his body functions and body characteristics (such as genetic codes and biometrics); privacy of behavior and action, which is associated with the individual's rights to maintain the confidentiality of their habits, political activities, and religious practices; privacy of personal communication, which includes any interception of any person's forms of communication, such as mail or phone calls, either through recording or interception; privacy of data and images, which includes any information and images, or videos related to the person; privacy of thoughts and feelings, which addresses the balance of power between the state and the person; privacy of location and space; and, lastly, privacy of association [16]. Nevertheless, privacy concerns in IoT environments extend beyond just the users who are actively connected to the internet; it may also impact individuals who are present in the IoT environment but are not using any IoT services. As internet applications constantly capture data flow, it becomes crucial to implement a reliable authentication procedure to prevent potential privacy violations. However, achieving this in an IoT setting can be challenging, and it is important for IoT environments to maintain individuals' privacy. Any data collected must only be used for its intended purpose and stored only for as long as necessary.

9.5 GAN FRAMEWORK

In recent years, machine learning has emerged as a crucial aspect of computer science development and data generation. It has become an integral part of the artificial intelligence (AI) community's efforts to enable machines to comprehend the complexities of the world in the same way humans do. As machine learning algorithms rely on data representation, scholars have proposed learning approaches to improve data generation [17]. Deep learning is one such approach used to augment data representations for machine learning algorithms. This approach involves creating simple representations that can extract high-level abstract features that surpass those extracted by other methods [18]. These learning approaches can be categorized into two groups: supervised learning with labeled datasets and unsupervised learning with unlabeled datasets. Generative models are a class of technology that employs unsupervised learning. These models apply a range of techniques, including Markov chains, maximum likelihood, and approximate inference for data generation [19].

There are three main categories of generative models, namely, GANs, variational autoencoder (VAE), and autoregressive networks. VAEs are probabilistic graphical models that aim to depict the probability distribution of data, leading to generating samples that tend to be noisier than those generated by GANs. On the contrary, autoregressive networks such as PixelRNN tackle image generation by predicting and generating pixels one by one, which makes the process slower compared to GANs that can process samples all at once. GANs, as a type of probabilistic generative model, utilize an internal adversarial training mechanism that works well even without knowing the density of probability [19].

The initial generative models faced limitations when it came to the generalization process. In 2014, Goodfellow proposed a new generative model, which addressed these issues. The GAN is an example of a generative model that takes a training dataset with samples from a distribution and learns to estimate that distribution by creating a probability distribution model. Some models explicitly estimate the distribution while others only generate samples from it, and some models can do both. The fundamental concept of GANs involves two networks: the generator and the discriminator. The generator creates samples that come from the same distribution as the training data, while the discriminator distinguishes between real and fake samples and learns using supervised learning techniques by classifying the inputs into either real or fake [10]. The framework of GANs is composed of two deep neural networks: the discriminator, which can be thought of as a policeman and the generator, which can be thought of as a counterfeiter attempting to create fake money. Each of these networks is represented by a different function controlled by a set of parameters. The generator is trained to deceive the discriminator, much like the counterfeiter attempts to fool the policeman into accepting their fake money.

Training examples are randomly sampled from the training set (x) and used as inputs for the discriminator function D. The discriminator produces a probability output that describes whether the input is real (with a probability close to 1) or fake (with a probability close to 0), assuming that half of the inputs are fake and the other half are real. The discriminator's objective is to produce probabilities that its inputs are real ($D(x)$ close to 1). The GAN framework uses different loss functions to estimate the loss of the discriminator and generator. The generator function may vary across different frameworks, but the discriminator functions remain the same. One function that can be employed to compute the loss of the generator is the zero-sum game, where the loss of both the generator and discriminator within any given framework is equivalent to zero, leading to the following result:

$$J^{(G)} = -J^{(D)}. \tag{9.1}$$

An alternative formula used to estimate the generator's loss in non-zero-sum frameworks involves a method that employs cross-entropy reduction for the generator. Rather than inverting the cost of the discriminator to determine the generator's loss, the target utilized to construct the cross-entropy loss is inverted instead. The resulting loss for the generator can then be obtained as:

$$J^{(G)} = -\frac{1}{2}\sum_{z} \log D(G(z)). \tag{9.2}$$

Frameworks with zero-sum costs are often referred to as min-max, as they involve minimizing in the outer loop and maximizing in the inner loop. This approach has been extensively employed to demonstrate the learning capability of GANs.

9.6 RESEARCH OBJECTIVES

The aim of this chapter is to explore a novel mechanism that utilizes GANs to develop a privacy-preserving solution for IoT data. The proposed approach is designed to minimize the likelihood of breaching the original data or any portion of it. There are several related studies to our approach, and readers are encouraged to read the works in Ref. [7,20,21].

The exponential growth of IoT ecosystems has given rise to a significant surge in the generation and exchange of sensitive data between different devices making it challenging to maintain individuals' privacy. Furthermore, most publicly available datasets used by researchers are insufficiently anonymized and can reveal sensitive information about individuals. Despite the existing research efforts on privacy-preserving data mining techniques, the use of GANs for privacy preservation is still relatively nascent. To address the primary research objectives outlined in the previous section, we have formulated the following research questions:

- **RQ1:** How effective is the use of generative adversarial networks (GANs) in generating anonymized datasets from the original IoT data?
- **RQ2:** What is the level of accuracy achieved in the generated datasets compared to the original dataset? To address this question, we conducted experiments to generate anonymized datasets using GANs and evaluated our proposed model. These experiments aimed to determine whether our approach provides better preservation of privacy compared to other existing methods. We measured both privacy and utility loss to assess the effectiveness of our method.

9.6.1 LIMITATION OF THE SCOPE

GAN has recently attracted researchers due to its performance as a generative model. This model is used to estimate the underlying distribution of a dataset and randomly generate realistic samples according to their estimated distribution. This threatens the data's privacy if exposed to this type of processing, as it will lead to data disclosure of many sensitive data related to individuals. Due to all the above, we are going to apply a privacy-preservation mechanism using GAN. Moreover, it has become a promising approach that enables analysts to generate an unlimited amount of synthetic data for analysis tasks without disclosing the privacy of training data.

Our GAN model is designed to handle two class labels at a time when generating datasets, which can be challenging for datasets that feature more than two class labels. Therefore, datasets containing multiple class labels need to be transformed into binary classes, resulting in a less accurate distribution for any of the classes compared to the distribution of the whole dataset.

9.7 LITERATURE REVIEW

9.7.1 DATA ANONYMIZING

Various methods have been proposed to ensure the privacy of static data, which is not subject to the same challenges faced by streaming data. This can be referred to as the need for incremental analysis of data streams, the limited time and buffer space used, and the ever-evolving nature of the data streams in general. In Ref. [22], the authors tried to take advantage of tracking the correlation and autocorrelation structure of the multivariate data stream by introducing an algorithm that is aiming to provide a prominent level of privacy. They created their approach by adding random perturbation to the data stream, which preserves its statistical properties.

The study by Alcaide et al. [23] created an approach to enhance the privacy elements along with the security properties of the IoT data. They focused on the anonymity of the IoT users and the unlinkability of their interactions during dealing with physical objects in the cyber community. They were motivated to guarantee that the use of sensitive information will not be violated and disclosed. The authors provided a fully decentralized anonymous authentication protocol to design a framework of decentralized founding nodes that present the participant's role in the ad-hoc community. This type of model is self-adaptive and target-driven, which adapts to the changes in the communities' behavior and serves common interests at both small and large communities. They have used the participants' indications and application feedback to promote behavioral changes in IoT communities. The main feature of this model is decentralization, so it is appropriate to be applied to any type of organization and to be set up based on the organizations' rules and regulations.

The authors in Ref. [24] relied on the k-anonymity principle and introduced a privacy-preserving algorithm to protect users' locations. Their proposed algorithm is implemented for the secure use of location-based services based on the entropy metric and the obtained information. The enhanced version of the dummy-location selection algorithm works on selecting a dummy location based on enlarging the cloaking region to guarantee the separation of the locations as far as possible.

The study by Wang et al. [25] evaluates attribute-based encryption (ABE) as a privacy-preserving solution. ABE is considered a type of public-key encryption. It is characterized by allowing access control and flexibility of the keys' exchange. The assessment procedures include network delay, control overhead data, CPU process, memory, power usage, and execution time.

The traditional solution for preserving data privacy is based on hop-by-hop or end-to-end encryption, which is now not efficient for protecting the data from any external threat. The authors in Ref. [26] proposed an encryption schema for securing the collaboration of IoT objects by homomorphic encryption that combines the fully additive encryption with the fully additive secret sharing to fulfill the required properties. Their proposed schema is built based on the elliptic curve cryptography augmented with threshold secret sharing to ensure the confidentiality and integrity of the collected data. This solution focuses on securing the part related to collecting the data from the IoT objects to use it by a variety of applications connected to the IoT environment. The solution considers the case where the aggregation function executed by the gateway relies on the additional operator over multiple variables. The major challenge for the gateway in this scenario is to perform operations on the aggregated information without breaking the confidentiality enforced by the source encryption at the device level. This allows the combination of two interesting properties, which are privacy preservation and cost-efficiency (in terms of bandwidth).

Different privacy models have been proposed for protecting big data, including static data and data streams. Using statistical disclosure control methods (SDC), these models can be enforced to protect private data against re-identification disclosure. The authors in Ref. [27] provided a primitive called steered micro aggregation that is generated to be a general mechanism and considers the emerging constraints. The proposed approach can handle and process both types of big data by adding weighted and proper attributes which are designed in specific ways. They also showed how t-closeness for static data and k-anonymity for data streams could be achieved by adding these artificial attributes using various models. The experimental results of t-closeness enforcement showed sensitive data controlling within-cluster variability. This mechanism also allowed other privacy models to be satisfied under specific conditions such as differential privacy.

It is known that IoT devices and transactions are subject to violations by several attacks (i.e., tracking and profiling). The authors in Ref. [28] developed a privacy-preserving method in intelligent transportation systems depending on the game theory between data holder (i.e., driver and intelligent devices) and data requester DR (i.e., employer and supplier). The new technique considers environmental properties such as power, memory, and communication regarding the suitable level of cryptographic mechanisms. The set of cases is defined for DH and DR and used the Markovian chain to model the transitions. A utility function is used to assess the tradeoff between the privacy characteristics to decide whether there is a disclosure of the sensitive data or not.

The study of Ref. [29] developed a privacy-preserving solution for IoT healthcare big data systems using access control techniques. The main objectives are obtaining secure healthcare data in two scenarios (i.e., normal and emergency) along with smart de-duplication of the big data system. To view sensitive patients' data over the IoT communication network, the proposed solution encrypts data and sends it to the storage system. Healthcare staff, in several levels, are granted an access control policy to manage access to information. The classical access control methods allow authorized users to get the information but prevent first-aid treatment in emergency cases. This study adopted a new method consisting of two access control levels; for normal cases, secret keys will be employed to decrypt data. While in emergency scenarios, the patient's data will be retrieved by a password-based break-glass access method. Also, the proposed method reduces the required memory and removes the duplicate patient files by adopting a secure de-duplication technique. The evaluation results showed the high-security performance and effective feasibility of applying the proposed solution. Their work, on the contrary, includes checking data integrity remotely which implies a trusted third party to oversee the outsourced data's integrity, which may be invalid in practice.

Zhao et al. [30] have proposed a schema that dispenses with the third party to check data integrity. They have utilized a blockchain architecture to construct a scheme for privacy-preserving remote data integrity checking in IoT information systems without TTP. This scheme leverages the lifted EC-ElGamal cryptosystem and the blockchain to support efficient public batch signature verifications and protect the IoT systems' security and data privacy. This scheme is more suitable for practical applications in the data management systems without involving and resisting the leakage of data privacy caused by the third party.

9.7.2 AUTHENTICATION AND AUTHORIZATION

Wireless sensor network (WSN) technology has been widely used in recent years in several applications. To obtain the required security/privacy levels, several WSN mutual authentication and exchange key protocols were proposed, and numerous studies focused on the vulnerabilities of these protocols [31]. The study in Ref. [32] conducted by He et al. highlighted several attacks for the existing protocols, such as offline password guessing attacks, user/sensor node impersonation attacks, and modification attacks. The authors presented a novel WSN temporal-credential-based mutual authentication and key management approach. The assessment results showed that this can be a new promising approach as it is an efficient solution for power consuming and provides an acceptable security level for WSNs. The study conducted in Ref. [31] also presents a high performance and new user authentication and a key management approach for WSN that is integrated into IoT. In their proposed approach, users will authenticate a particular sensor node from heterogeneous WSN without calling the gateway node. Their approach depends on symmetric cryptographic mechanisms and provides high efficiency.

The study of Ref. [33] shows the continuous efforts for enhancing the security properties and efficient methods for combining wireless sensor networks (WSN) with IoT. In Ref. [33], Farash et al. found that the approach by Ref. [31] is vulnerable to some attacks. So, this study proposes an enhanced security approach. The authors in Ref. [31] proposed a new method for privacy-preserving with novel functionalities. The evaluation security results conducted using BAN-logic and AVISPA simulation engines showed a remarkable level of security that prevents WSN attacks. Performance comparisons yielded more hash computation costs done by the user or at the gateway node. Moreover, the proposed approach required less storage consumption and gave dynamic growth of WSN without affecting the functionality or the authentication process.

Privacy-preserving and efficiency are the major challenges of smart homes. Due to the nature of communication between the components at smart homes via wireless networks, the personal data can be exploited by attackers who can simply spy on the data traffic between these components and, as a result, reduce their daily activities. Considering these challenges, the authors of Ref. [34] proposed a work based on certification authority (CA) as an authentication protocol for the sensors. Their work discusses that all smart home entities should own a special CA and pass the privilege check to manipulate the other entities. This type of smart home design relies on a trusted CA center, which has the privilege of assigning a newly registered entity. Sensors are an integral component of smart home architecture, but their limited computing capabilities render them inadequate for executing complex computational processes. The emerging EU law has identified several regulations concerning users' privacy provision. These laws provided that IoT users must consent to the services that can interact with them and their data. To realize this, Kirkham et al. [35] suggested extra data processing at the data collection stage. They explained the benefits of using hybrid IoT data processing solutions as they are made for data filtering and can be invested in privacy and service provision. They showed that the solution has many networking benefits processing with the cloud, reducing the load produced from extra processing places, and providing additional capacity from the cloud. As a result, it reduces many of the problems that might appear, such as latency that will be added to the sensor board's operations and the bottleneck problem for the quality of services.

Protecting user privacy in the IoT environment poses a significant challenge with IoT sensors and devices collecting a massive amount of personal information about users' daily life without their consent. While service providers frequently require users to explicitly define their privacy preferences, the limitation of the available time to read consents, the lack of motivation, the cognitive burden of users, and the restricted IoT user interface prevent users from making their privacy decisions regarding IoT services. To mitigate these issues, researchers have leveraged intelligent systems equipped with machine learning to predict users' privacy decisions in the IoT environment and assist users in having suitable privacy preferences. For example, a study by the authors of Ref. [36] tested 172 users in a simulated IoT and collected the users' opinions regarding their preferences about hypothetical personal information tracking cases. The authors employed the K-modes clustering technique on the users' answers. The obtained results indicate four distinguished clusters; in each of these clusters, a relationship between IoT context and user attitudes was detected. Authors applied to learn conditional inference trees to predict users' preferences, and the model yielded a prediction accuracy of 77% using a 10-fold cross-validation method.

The implementation of access control in an IoT environment faces two primary challenges: the constrained nature of the smart objects and the risk of compromising user privacy when involving third-party entities in the access control process.

Ouaddah et al. [37] proposed a new decentralized pseudonymous and privacy-preserving authorization management framework that can deal with IoT distributed nature using fair access. Fair access is a distributed privacy-preserving access control framework introduced as a balance solution for solving the problem of centralized and decentralized access control that are using the cryptocurrency blockchain mechanism and allowing the users to identify and control their privacy level. In this proposed framework, the authorization decisions are made by using a smart contract that leans on policies of contextual access control, as well as being used for fine-grained expression.

The enormous expansion of IoT-connected devices that collect sensitive information about users and the environment around them makes it necessary for manufacturers to prepare and implement their Privacy Policy Agreement (PPA) for all their respective devices. These policies include expressing what kind of information the device collects, where these data are stored, and for what reason it might be used. In the study conducted in Ref. [2], the authors showed that half of the manufacturers to which the study was applied do not have a firm and specific privacy policy for their IoT devices. They also tested the data transition between IoT devices and the cloud and proved using specific privacy criteria that the communication between these devices does not comply with what they are stated in their PPA.

In the same context, end-to-end authentication is considered a main issue in IoT environments due to the technical differences between IoT devices and controllers. In literature, we have several authentication protocols to guarantee security and privacy needs. The study in Ref. [38] presents an anonymous end-to-end mutual authentication and key management approach depending on the ZigBee protocol. ZigBee allows high-level communication by establishing personal networks that require low power and digital radios suitable for smart home networks. The new protocol employs secret key encryption to end-to-end IoT authentication process to achieve confidentiality. The proposed method updates the session keys for each new communication to obtain the anonymity, untraceability, and unlinkability of IoT devices. It also uses HMAC technique based on a trusted incremental counter to obtain the data integrity. The security is validated by BAN logic and assessed by automated validation of internet security protocols and applications (AVISPA) toolset, which indicates the safety of the proposed method.

9.7.3 EDGE COMPUTING AND PLUG-IN ARCHITECTURE

Several studies were conducted to address protecting users' data privacy by investigating on-device sensor abstractions for augmented reality applications and therefore preventing these data from accidental leakage [39]. The efforts of the study come to address devices and applications that have privileged access to raw sensor data [39].

In the area of data privacy protection, Al-Hasnawi et al. [40] introduced a solution to protect IoT users' sensitive data, which contains personally identifiable information (PII) in its life cycle. The policy enforcement fog module is a proposed module running to enforce privacy policies on sensitive data collected from different sources. Their proposed software module uses the power of policy enforcement at the edge-fog infrastructure. Their scheme's main idea is to protect the data from the moment it originated through the sensor until data destruction. This module deals with two kinds of applications for data processing; the first one is the local and real-time IoT applications, while the second one is remote fogs, clouds, and non-real-time applications. The authors present a data privacy solution for each one of these applications. The solution uses policy enforcement directly on the collected IoT data for local fog nodes closest to the data source. While the second type of applications' problem is solved by using the active data bundle (ADB), a self-protecting construct for remote applications. Their approach works in parallel with protecting data and policy enforcement engine. It is created initially at the source node and automatically enabled at the destination nodes [41].

Ant colony optimization approach is proposed by Lin et al. [42]. This approach uses transaction deletion to secure confidential and sensitive information. Each ant represents a set of possible deletion transactions used for hiding sensitive information and securing the users' identity. To assist the reduction of dataset scans in the evolution process, the authors utilized the pre-large concept, and to find the optimized solution, they adopted external solutions that discovered Pareto solutions.

9.7.4 USING GENERATIVE ADVERSARIAL NETWORK (GAN) IN PRIVACY DATA ANALYTICS

In the massive growth of the data generated from multiple sources every second in IoT environment, adversaries are mining private information for potential benefits. To target this concern, Qu et al. [20] proposed a GAN-driven noise generation method under a differential privacy framework. By adding a new perception of differential privacy identifier, the generator is forced to produce a differential private noise and then the discriminator with the identifier gaming to derive the Nash equilibrium.

GANs and their variants are highly effective data generation models that can produce large amounts of high-quality data by learning the complex semantics of the underlying data distribution from training data. However, the rise of memorizing sensitive information from the training dataset and subsequently generating data that pose a privacy risk remains a challenge for these models. Authors in Ref. [43] proposed PPGAN, a privacy-preserving GAN, model that achieves differential privacy in GAN by adding a well-designed noise to the gradient during the model learning stage. To improve the stability and compatibility of the model, a Moments Accountant strategy is proposed and working on controlling the privacy loss.

A differentially private GANobfuscator was proposed by Ref. [7] to ensure that critical information will not divulge when generating sensitive information. This was achieved by adding a designed noise to gradients during the learning process. This generative model is generating an unlimited amount of synthetic data for analysis purposes and helps in preserving the privacy of the training data. The authors in this study also developed a gradient-pruning strategy to improve the scalability and stability of the data training when using the GANobfuscator.

In the same context, the authors in Ref. [44] discussed the vulnerability in IoT devices caused by increased network traffic and the expansion of IoT devices. They proposed a robust approach based on deep Q-network (DQN) and GAN for unsolicited proxy detection in IoT traffic. This approach is designed for discerning anomalies and preventing unauthorized access to users' information by an adversary and can withstand malicious alterations of connection information. Their proposed approach utilized the GAN discriminator as a target network for DQN to detect the proxied requests/responses in the connection information.

Moreover, Wang et al. [21] proposed PART-GAN, a practical privacy-preserving generative model for time series data augmentation and sharing. This model enables the local data curator to provide a freely accessible public generative model from the original data for cloud storage. This approach's key advantages allow for generating an unlimited amount of time series data under the condition of addressing incomplete and temporal irregularity issues and with a given classification label. The suggested approach provides a differential private time series data augmentation and sharing technique. In addition, and by applying the optimization strategies, the authors tried to address the trade-offs between utility and privacy.

9.8 OVERALL RESEARCH DESIGN

Figure 9.1 illustrates the overall conceptual design of our proposed solution. As shown in the figure, in phase 1, the original data are generated from IoT devices and represented in a dataset. Along with each record, we represent each record's activity label. Generally, we used the accelerations and angular velocity values generated from smartphone sensors (gyroscope and accelerometer); these values can be processed and then used to learn more about users' status [45]. In phase 2, we transform the numerical representation of the acceleration and angular velocity values into 0 or 1 representations. In the third phase, the dataset is generated, and the noise is added using GAN. Finally, the generated dataset is released in order to be used in data analytics and classification.

Data Samples: To generate a private dataset and evaluate our generative model, we used the HAPT2015 dataset from Smart Lab, an Italian research laboratory for computational intelligence and data analytics.

The HAPT2015 dataset is an activity recognition dataset consisting of recordings of 30 subjects with ages ranging between 19 and 48 years [46]. These subjects were asked to perform a protocol of six basic activities: three static postures (standing,

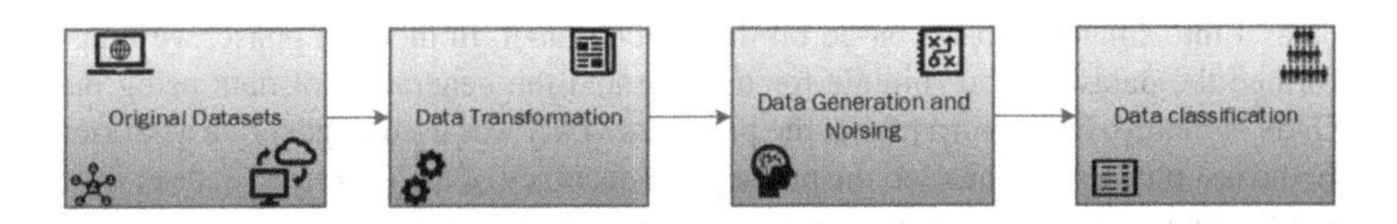

Figure 9.1 Overall research design.

Table 9.1
Characteristics of HAPT2015 Dataset

User Number ID	Feature Vectors	Activity Number ID
Consist of numerical numbers between 1 and 30, and these numbers represent the identifiers of the subjects who did the experiment	A 561-feature vector with time and frequency domain variables. Features are normalized and bounded within [−1, 1]. Each feature vector is a row on the 'X' and 'y' files. The units used for the accelerations (total and body) are 'g's (gravity of earth → 9.80665 m/seg^2). The gyroscope units are rad/seg	Identifiers of activity labels represented with numbers associated with it: (Walking) (Walking upstairs) (Walking downstairs) (Sitting) (Standing) (Laying) (Stand to Sit) 8 (Sit to Stand) 9 (Sit to Lie) 10 (Lie to Sit) 11 (Stand to Lie) 12 (Lie to Stand)

sitting, lying) and three dynamic activities (walking, walking upstairs, and walking downstairs), in addition to postural transitions that occurred between the static postures (Stand-to-Sit, Sit-to-Stand, Sit-to-Lie, Lie-to-Sit, Stand-to-Lie, Lie-to-Stand). These activities are recorded while the subjects are carrying a waist-mounted smartphone with embedded inertial sensors during the experiment execution. The embedded accelerator and gyroscope of the device captured 3-axial linear acceleration and 3-axial angular velocity at a constant rate of 50 Hz. The sensor signals that are generated from (accelerometer and gyroscope) were pre-processed. The preprocessing was performed by applying noise filters. Later, these signals were sampled in fixed-width sliding windows of 2.56 seconds and 50% overlap (128 readings/window). Each of these windows represents a vector of 561 features. In this research, we used the resulting dataset that includes records of activity windows. Each record is composed of the attributes as shown in Table 9.1: The subject ID, feature vectors, and the associated activity label. We extracted the first 100 feature vectors and their associated activity label for each record and used them in our experiments.

9.9 METHODOLOGY

We used a labeled dataset in the data preparation phase, and then, we divided the dataset into smaller subsets based on the activity label. In the next phase, we transformed the dataset to be suitable for training and the generation of data using our GAN structure. In the third phase, the proposed GAN was used to generate a dataset. In the last phase, we evaluated the privacy and accuracy of the generated dataset from GAN using differential entropy compared to the original dataset.

9.9.1 DATA PREPARATION

The dataset of HAPT2015 consists of two splits: the first is the training dataset and the second is the testing one. Nevertheless, since using the original dataset is not suitable for our experiments, the training dataset with the testing dataset was merged into one dataset of 10,929 records consisting of an attribute identifier for the subjects who performed the activity. The total number of features was 103 feature vectors along with their associated activity labels. The dataset was divided into six subsets; each one consists of the records that are associated with two distinct activity labels as shown in Table 9.2.

It was essential to transform the feature vector values for the 103 attributes to be 0 or 1. The values that are ranged in the period $[-1, 0)$ are replaced with 0, and the values that are ranged within the period $[0, 1]$ are replaced with 1. The activity labels are also masked with 0 and 1 values in each dataset. Table 9.3 represents the original values and the transformed value for each dataset.

Dataset *Generating with Noise Addition:* In this phase we used GAN implementation to generate a noisy version of the dataset from the original one. The implementation of GAN supports both CUDA and CPU. In this research, the implementation was built using CPU only. Figure 9.2 simplifies the structure of GAN to generate a private dataset; the first part of the GAN represents the generator that is used to generate fake records after noise addition. The generator is trained to generate noisy records similar to the original records but are not easily distinguished by the discriminator.

The second part of GAN is the discriminator, which is the second neural network where they are responsible for distinguishing the adversarial records (i.e., fake records) from the benign (i.e., real records). The generated dataset is built according to the feedback sent to the generator each time the discriminator labels the generated record as fake. The discriminator is fed with feedback at each training round where the record is labeled as fake. This generation and distinguishing process is performed until the last epoch of training.

Table 9.2
Dataset with Two Activity Labels

Dataset Name	Activity Labels	Number of Records
DS-1	(1, 2)	3,266
DS-2	(3, 4)	3,207
DS-3	(5, 6)	3,935
DS-4	(7, 8)	102
DS-5	(9, 10)	191
DS-6	(11, 12)	222

Table 9.3
Dataset Value Transformation into 0–1

Dataset Name	Feature Vector (Original Value x)	Feature Vector (Transformed Values)	Activity Label (Original Values)	Activity Label (Transformed Values)
DS-1	$-1 \le x < 0$	0	1	0
	$0 \le x \le 1$	1	2	1
DS-2	$-1 \le x < 0$	0	3	0
	$0 \le x \le 1$	1	4	1
DS-3	$-1 \le x < 0$	0	5	0
	$0 \le x \le 1$	1	6	1
DS-4	$-1 \le x < 0$	0	7	0
	$0 \le x \le 1$	1	8	1
DS-5	$-1 \le x < 0$	0	9	0
	$0 \le x \le 1$	1	10	1
DS-6	$-1 \le x < 0$	0	11	0
	$0 \le x \le 1$	1	12	1

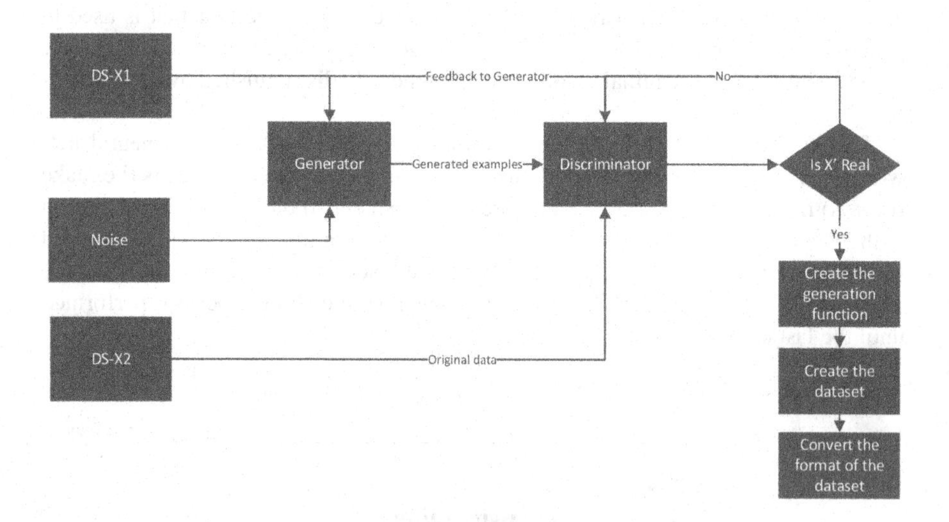

Figure 9.2 Data generation framework.

9.10 DATA ANALYSIS AND INTERPRETATION

In this research, several evaluation metrics are used to test the accuracy and privacy of the proposed solution. For testing the accuracy, we utilized several accuracy measures to validate the effectiveness of the data generation model such as TP rate, FP rate, Precision, Recall, F-measure, classification accuracy.

9.10.1 PRIVACY MEASURES

For measuring privacy, we utilized differential entropy to measure the privacy of generated data. Conditional privacy is an average measure of privacy. It was proposed in the context of distribution reconstruction after additive perturbation. This measure is based on the differential entropy of random variables. The differential entropy of A given $B = b$ is:

$$h(A|B) = -\int_{\Omega_{A,B}} f_{A,B}(a,b) \log_2 f_{A|B=b}(a)\, da\, db \tag{9.3}$$

where A is a random variable that describes the data, and B is the variable that gives information on $A.\Omega_{A,B}$ and identifies the domain of A and B. Therefore, the average conditional privacy of A given B is:

$$\prod(A|B) = 2^{h(A|B)} \tag{9.4}$$

9.10.2 ACCURACY MEASURES

The purpose of this measure is to calculate the classification accuracy (ACC) of each ML algorithm, and it is calculated as:

$$\text{ACC} = \frac{(\text{TP} + \text{TN})}{P + N} \tag{9.5}$$

9.10.3 INCORRECT CLASSIFICATION

The purpose of this measure is to find out the misclassification rate of each of the classes, and it is calculated as:

$$\text{Misclassification rate (MR)} = 1 - \text{classification accuracy} \tag{9.6}$$

Precision:
– Precision of original vs. generated data = $\frac{\text{TP}}{\text{TP+FP}}$
(11)

9.10.4 F-MEASURE

We used this measure to find out the harmonic means of precision and recall measures.

$$F1 = \frac{2\text{TP}}{(2 + \text{TP} + \text{FP} + \text{FN})} \tag{9.7}$$

9.10.5 PRIVACY

Differential entropy is an average measure of privacy initially proposed in the context of distribution reconstruction after additive perturbation. The results of the dataset's privacy are evaluated with a fixed value of the latent vector and batch size, and the number of the epoch is changed along the dataset generated.

9.10.6 PRIVACY RESULTS USING DIFFERENT NUMBER OF EPOCHS

To evaluate our method in terms of privacy, the differential entropy of the generated datasets is calculated. The experiments run on three subsets. Table 9.4 provides information about each subset.

Each experiment has generated 19 datasets with the same values of latent vector and batch, but with a different number of epoch values varying between 10 and 1,000. Table 9.5 presents the privacy values for the dataset generated from the three experiments.

Table 9.4
X1 and X2 Subsets

Experiment Number	X1	X2
1	DS-A	DS-1
2	DS-B	DS-2
3	DS-C	DS-3

Table 9.5
Privacy Values Datasets Generated from Experiments 1–3

Experiment #	# of Epochs	Z	Batch Size	Experiment 1	Experiment 2	Experiment 3
_.1	10	10	32	0.88	0.855	0.817
_.2	20	10	32	0.90	1.000	0.826
_.3	30	10	32	0.55	0.817	0.712
_.4	40	10	32	0.63	0.778	0.807
_.5	50	10	32	0.88	0.951	0.788
_.6	60	10	32	0.92	1.03	0.778
_.7	70	10	32	0.87	0.788	0.836
_.8	80	10	32	0.61	0.788	0.530
_.9	90	10	32	0.68	0.778	0.817
_.10	100	10	32	0.76	0.996	0.884
_.11	200	10	32	0.70	0.653	0.586
_.12	300	10	32	0.63	0.759	0.568
_.13	400	10	32	0.69	0.740	0.778
_.14	500	10	32	0.62	0.701	0.826
_.15	600	10	32	0.88	0.711	0.788
_.16	700	10	32	0.55	0.653	0.857
_.17	800	10	32	0.56	0.750	0.722
_.18	900	10	32	0.59	0.644	0.586
_.19	1,000	10	32	0.58	0.721	0.875

After this experiment, we again revisit each of our research questions.

RQ1: To what extent is the use of generative adversarial networks (GANs) suitable for generating anonymized datasets from the original data?

Figure 9.3 illustrates the privacy of the three experiments based on the number of epochs. The number of epochs was selected using the state of the art of the existing research. Privacy varies based on the number of epochs. DS-2 records achieve the highest privacy values among the three datasets. A high privacy value is also attained when the number of epochs were 60 and 100. The

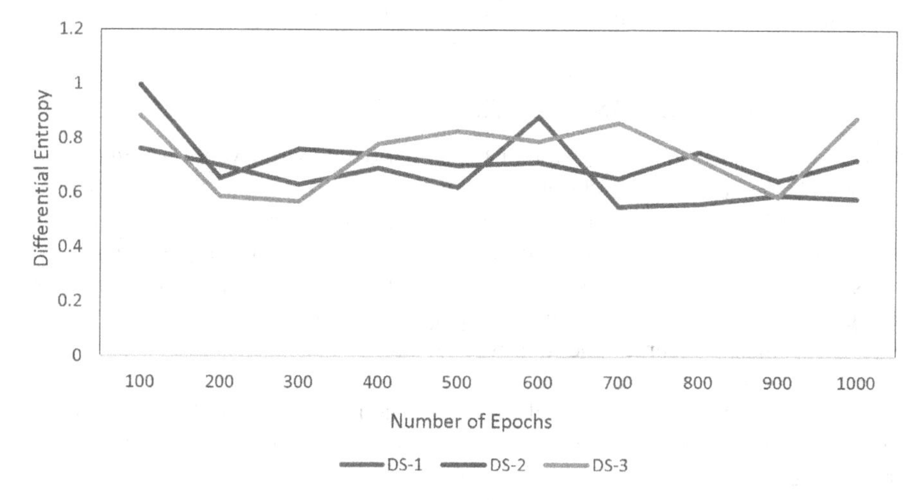

Figure 9.3 Privacy when varying the number of epochs.

Table 9.6
KNN Classification Measure for Original and Generated Datasets

KNN		Exp1	Exp2	Exp3	Exp4	Exp5	Exp6	Exp7	Exp8
	Time original	7.3	3.1	0.87	0.55	0.7	1.1	1.3	1.6
	Time generated	7.3	8.2	9.1	0.55	0.54	1	0.7	1.1
	CA original	0.87	0.99	0.99	0.93	0.95	0.91	0.99	0.99
	CA generated	0.81	0.99	0.93	0.98	0.97	0.90	0.98	0.99
	AUC original	0.94	1	1	0.96	1	0.97	0.99	1.0
	AUC generated	0.89	0.99	0.97	0.45	0.99	0.81	0.84	0.99
	F1 original	0.86	0.99	0.99	0.93	0.95	0.91	0.99	0.99
	F1 generated	0.81	0.99	0.92	0.97	0.97	0.88	0.97	0.99
	Prec. original	0.85	0.99	0.99	0.94	0.95	0.91	0.99	0.99
	Prec. generated	0.81	0.99	0.93	0.96	0.97	0.88	0.97	0.99
	Recall original	0.88	0.99	0.99	0.93	0.95	0.91	0.99	0.99
	Recall generated	0.81	0.99	0.92	0.98	0.97	0.9	0.98	0.99

best privacy value is attained for DS-2 which equals 1.03. The lowest privacy value attained was when the number of epochs equals 30 for DS-1. In addition, when the number of epochs was 80, the privacy value was 0.53 for DS-3, which is the lowest privacy value for that dataset. The results clearly indicate that there is a need for a fine-tuning mechanism to identify the best balance between privacy and utility.

RQ2: How much accuracy is achieved on the generated datasets?

For investigating the classification of generated dataset from original dataset, K-nearest neighbor classifier (KNN) is used. The results of the K-NN classification are shown in Table 9.6. The results clearly indicate that our method generates datasets that are comparable to the original dataset

9.11 CONCLUSION AND FUTURE WORK

The widespread use and integration of IoT in our daily lives have raised significant privacy challenges on various levels, and privacy-preserving methods have been proposed to address these challenges. In this research, a privacy-preserving GAN was introduced to generate synthetic data for IoT environments. Our proposed approach uses noise addition to achieve privacy while generating the dataset and then analyzes the generated datasets using machine learning algorithms such as KNN. The results showed promising accuracy and improved privacy.

For future work, we plan to investigate the performance of our method on a large-scale data set and explore more advanced techniques to enhance the privacy level of the generated datasets. We also aim to compare our method with other privacy-preserving techniques and evaluate its performance in various IoT applications. Furthermore, we will explore the possibility of incorporating differential privacy techniques into GANs to improve the privacy level of the generated datasets. Overall, this research provides a promising direction for preserving privacy in IoT environments and can be extended to various applications that require data privacy.

REFERENCES

1. I. Lee and K. Lee, "The Internet of Things (IoT): Applications, investments, and challenges for enterprises," *Business Horizons,* vol. 58, no. 4, pp. 431–440, 2015.
2. A. Subahi and G. Theodorakopoulos, "Ensuring compliance of IoT devices with their privacy policy agreement," in *2018 IEEE 6th International Conference on Future Internet of Things and Cloud (FiCloud)*, 2018, Barcelona, IEEE, pp. 100–107.
3. J. Gubbi, R. Buyya, S. Marusic, and M. Palaniswami, "Internet of Things (IoT): A vision, architectural elements, and future directions," *Future Generation Computer Systems,* vol. 29, no. 7, pp. 1645–1660, 2013.
4. A. Dorri, S. S. Kanhere, R. Jurdak, and P. Gauravaram, "Blockchain for IoT security and privacy: The case study of a smart home," in *2017 IEEE International Conference on Pervasive Computing and Communications Workshops (PerCom Workshops)*, 2017: IEEE, Kona, HI, pp. 618–623.

5. T. Pasquier, J. Singh, J. Powles, D. Eyers, M. Seltzer, and J. Bacon, "Data provenance to audit compliance with privacy policy in the Internet of Things," *Personal and Ubiquitous Computing,* vol. 22, pp. 333–344, 2018.
6. M. A. P. Chamikara, P. Bertók, D. Liu, S. Camtepe, and I. Khalil, "Efficient data perturbation for privacy preserving and accurate data stream mining," *Pervasive and Mobile Computing,* vol. 48, pp. 1–19, 2018.
7. C. Xu, J. Ren, D. Zhang, Y. Zhang, Z. Qin, and K. Ren, "GANobfuscator: Mitigating information leakage under GAN via differential privacy," *IEEE Transactions on Information Forensics and Security,* vol. 14, no. 9, pp. 2358–2371, 2019.
8. S. Arora, R. Ge, Y. Liang, T. Ma, and Y. Zhang, "Generalization and equilibrium in generative adversarial nets (gans)," in *International Conference on Machine Learning*, 2017: PMLR, pp. 224–232.
9. H. Bae, D. Jung, H.-S. Choi, and S. Yoon, "AnomiGAN: Generative adversarial networks for anonymizing private medical data," in *Pacific Symposium on Biocomputing 2020*, Fairmont Orchid,World Scientific, pp. 563–574.
10. I. J. Goodfellow, J. Shlens, and C. Szegedy, "Explaining and harnessing adversarial examples," arXiv preprint arXiv:1412.6572, 2014.
11. T. Salimans, I. Goodfellow, W. Zaremba, V. Cheung, A. Radford, and X. Chen, "Improved techniques for training gans," *Advances in Neural Information Processing Systems,* vol. 29, pp. 1–9, 2016.
12. M. Leo, F. Battisti, M. Carli, and A. Neri, "A federated architecture approach for Internet of Things security," in *2014 Euro Med Telco Conference (EMTC)*, Naples, 2014: IEEE, pp. 1–5.
13. A. Al-Fuqaha, M. Guizani, M. Mohammadi, M. Aledhari, and M. Ayyash, "Internet of Things: A survey on enabling technologies, protocols, and applications," *IEEE Communications Surveys & Tutorials,* vol. 17, no. 4, pp. 2347–2376, 2015.
14. J. Lin, W. Yu, N. Zhang, X. Yang, H. Zhang, and W. Zhao, "A survey on Internet of Things: Architecture, enabling technologies, security and privacy, and applications," *IEEE Internet of Things Journal,* vol. 4, no. 5, pp. 1125–1142, 2017.
15. R. Pateriya and S. Sharma, "The evolution of RFID security and privacy: A research survey," in *2011 International Conference on Communication Systems and Network Technologies*, Bangalore, 2011: IEEE, pp. 115–119.
16. S. Gutwirth, R. Leenes, P. De Hert, and Y. Poullet, *European Data Protection: Coming of Age*. Springer: Berlin, Germany, 2013.
17. E. Bengio, P.-L. Bacon, J. Pineau, and D. Precup, "Conditional computation in neural networks for faster models," arXiv preprint arXiv:1511.06297, 2015.
18. Y. LeCun, Y. Bengio, and G. Hinton, "Deep learning," *Nature,* vol. 521, no. 7553, pp. 436–444, 2015.
19. J. Yao, W. Pan, S. Ghosh, and F. Doshi-Velez, "Quality of uncertainty quantification for Bayesian neural network inference," arXiv preprint arXiv:1906.09686, 2019.
20. Y. Qu, S. Yu, J. Zhang, H. T. T. Binh, L. Gao, and W. Zhou, "GAN-DP: Generative adversarial net driven differentially privacy-preserving big data publishing," in *ICC 2019IEEE International Conference on Communications (*ICC), Shanghai, 2019: IEEE, pp. 1–6.
21. S. Wang, C. Rudolph, S. Nepal, M. Grobler, and S. Chen, "PART-GAN: Privacy-preserving time-series sharing," in *Artificial Neural Networks and Machine Learning–ICANN 2020: 29th International Conference on Artificial Neural Networks*, Bratislava, Slovakia, September 15–18, 2020, Proceedings, Part I 29, 2020: Springer, pp. 578–593.

22. M. A. P. Chamikara, P. Bertók, D. Liu, S. Camtepe, and I. Khalil, "An efficient and scalable privacy preserving algorithm for big data and data streams," *Computers & Security,* vol. 87, p. 101570, 2019.
23. A. Alcaide, E. Palomar, J. Montero-Castillo, and A. Ribagorda, "Anonymous authentication for privacy-preserving IoT target-driven applications," *Computers & Security,* vol. 37, pp. 111–123, 2013.
24. B. Niu, Q. Li, X. Zhu, G. Cao, and H. Li, "Achieving k-anonymity in privacy-aware location-based services," in *IEEE INFOCOM 2014-IEEE Conference on Computer Communications*, Toronto, 2014: IEEE, pp. 754–762.
25. X. Wang, J. Zhang, E. M. Schooler, and M. Ion, "Performance evaluation of attribute-based encryption: Toward data privacy in the IoT," in *2014 IEEE International Conference on Communications (ICC*), Sydney, 2014: IEEE, pp. 725–730.
26. J. Zouari, M. Hamdi, and T.-H. Kim, "A privacy-preserving homomorphic encryption scheme for the internet of things," in *2017 13th International Wireless Communications and Mobile Computing Conference (IWCMC),* Valencia, *2017: IEEE, pp. 1939*–1944.
27. J. Domingo-Ferrer and J. Soria-Comas, "Steered microaggregation: A unified primitive for anonymization of data sets and data streams," in *2017 IEEE International Conference on Data Mining Workshops (ICDMW),* New Orleans, LA, 2017: IEEE, pp. 995–1002.
28. A. R. Sfar, Y. Challal, P. Moyal, and E. Natalizio, "A game theoretic approach for privacy preserving model in IoT-based transportation," *IEEE Transactions on Intelligent Transportation Systems,* vol. 20, no. 12, pp. 4405–4414, 2019.
29. Y. Yang, X. Zheng, W. Guo, X. Liu, and V. Chang, "Privacy-preserving smart IoT-based healthcare big data storage and self-adaptive access control system," *Information Sciences,* vol. 479, pp. 567–592, 2019.
30. Q. Zhao, S. Chen, Z. Liu, T. Baker, and Y. Zhang, "Blockchain-based privacy-preserving remote data integrity checking scheme for IoT information systems," *Information Processing & Management,* vol. 57, no. 6, p. 102355, 2020.
31. M. Turkanović, B. Brumen, and M. Hölbl, "A novel user authentication and key agreement scheme for heterogeneous ad hoc wireless sensor networks, based on the Internet of Things notion," *Ad Hoc Networks,* vol. 20, pp. 96–112, 2014.
32. Y. He, P. Bahirat, B. P. Knijnenburg, and A. Menon, "A data-driven approach to designing for privacy in household IoT," *ACM Transactions on Interactive Intelligent Systems (TiiS),* vol. 10, no. 1, pp. 1–47, 2019.
33. M. S. Farash, M. Turkanović, S. Kumari, and M. Hölbl, "An efficient user authentication and key agreement scheme for heterogeneous wireless sensor network tailored for the Internet of Things environment," *Ad Hoc Networks,* vol. 36, pp. 152–176, 2016.
34. Y. Jie, J. Y. Pei, L. Jun, G. Yun, and X. Wei, "Smart home system based on IoT technologies," in *2013 International Conference on Computational and Information Sciences*, Ho Chi Minh City, 2013: IEEE, pp. 1789–1791.
35. T. Kirkham et al., "Privacy aware on-demand resource provisioning for IoT data processing," in *Internet of Things. IoT Infrastructures: Second International Summit, IoT 360° 2015*, Rome, Italy, October 27–29, 2015, Revised Selected Papers, Part II, 2016: Springer, pp. 87–95.
36. H. Lee and A. Kobsa, "Privacy preference modeling and prediction in a simulated campuswide IoT environment," in *2017 IEEE International Conference on Pervasive Computing and Communications (PerCom*), Kona, HI, 2017: IEEE, pp. 276–285.

37. A. Ouaddah, A. A. Elkalam, and A. A. Ouahman, "Towards a novel privacy-preserving access control model based on blockchain technology in IoT," in *Europe and MENA Co-operation Advances in Information and Communication Technologies*, 2017: Springer, pp. 523–533.
38. M. Alshahrani and I. Traore, "Secure mutual authentication and automated access control for IoT smart home using cumulative keyed-hash chain," *Journal of Information Security and Applications,* vol. 45, pp. 156–175, 2019.
39. J. Vilk et al., "SurroundWeb: Mitigating privacy concerns in a 3D web browser," in *2015 IEEE Symposium on Security and Privacy*, San Jose, CA, 2015: IEEE, pp. 431–446.
40. A. Al-Hasnawi, S. M. Carr, and A. Gupta, "Fog-based local and remote policy enforcement for preserving data privacy in the Internet of Things," *Internet of Things,* vol. 7, p. 100069, 2019.
41. A. Al-Hasnawi and L. Lilien, "Pushing data privacy control to the edge in IoT using policy enforcement fog module," in *Companion Proceedings of the10th International Conference on Utility and Cloud Computing*, 2017, pp. 145–150.
42. J. C.-W. Lin, G. Srivastava, Y. Zhang, Y. Djenouri, and M. Aloqaily, "Privacy-preserving multiobjective sanitization model in 6G IoT environments," *IEEE Internet of Things Journal,* vol. 8, no. 7, pp. 5340–5349, 2020.
43. Y. Liu, J. Peng, J. James, and Y. Wu, "PPGAN: Privacy-preserving generative adversarial network," in *2019 IEEE 25Th International Conference on Parallel and Distributed Systems (ICPADS)*, 2019: IEEE, pp. 985–989.
44. O. Kayode and A. S. Tosun, "Deep Q-network for enhanced data privacy and security of IoT traffic," in *2020 IEEE 6th World Forum on Internet of Things (WF-IoT)*, 2020: IEEE, pp. 1–6.
45. D. Anguita, A. Ghio, L. Oneto, X. Parra, and J. L. Reyes-Ortiz, "Human activity recognition on smartphones using a multiclass hardware-friendly support vector machine," in *Ambient Assisted Living and Home Care: 4th International Workshop, IWAAL 2012*, Vitoria-Gasteiz, Spain, December 3–5, 2012. Proceedings 4, 2012: Springer, pp. 216–223.
46. J.-L. Reyes-Ortiz, L. Oneto, A. Samà, X. Parra, and D. Anguita, "Transition-aware human activity recognition using smartphones," *Neurocomputing,* vol. 171, pp. 754–767, 2016.

[illegible]

[illegible] M. A[illegible] [illegible]

[illegible]

[illegible] 150.

[illegible]

Index

Note: **Bold** page numbers refer to tables and *italic* page numbers refer to figures.

advanced evasion techniques 24
advanced metering infrastructure (AMI) system
  asset classifications **10**
  benefits 1
  communication layer vulnerabilities 13, **14**
  communication networks 4
  components 4–5, *5*
  data layer vulnerabilities 12–13, **13**
  definition 3
  hardware layer vulnerabilities 11–12, **12**
  implementation 9
  information security risk assessment 6–9, *7*
  primary goal 2, 3–4
  risk assessment 3
  risk identification 9–11
  risk profiling phase for 14–16, **15,** *16,* **17**
  risk treatment phase for 16, 18
  stages 5–6
adversarial machine learning 30–31
adversary user (AU) 58, *58,* 59
air transport 204
AMI system *see* advanced metering infrastructure (AMI) system
android-based mobile devices 24
Android Package Kit (APK) 34, 36
annualized loss expectancy (ALE) 8
annualized rate of occurrence (ARO) 8
ant colony optimization approach 220
artificial intelligence (AI) 18–19, 26
attribute-based access control (ABAC) framework 154
attribute-based encryption (ABE) 215–216
authentication 101
  smart homes privacy and security 160
  smart locks 150
authorisation 90, 101–102, 116
  generative adversarial networks 217–219
  smart locks 150
autoencoders; *see also* restricted Boltzmann machines (RBMs) model
  cyber-physical systems using 193, *193, 194*
  fraud and anomaly detection 196–199, *197,* **198**
  targeted problems using 190–191
automated meter reading (AMR) system 4
automation devices 81
automation security 147, 166

big data security 203, 209, 216
biometrics 149

catastrophic level, of attack 16
CCA-secure secret key encryption 155
certain level, of attack 15
certification authority (CA) 218
CNN *see* convolutional neural network (CNN)
cloud computing technology 18
cognitive communication systems 167
companion applications vulnerabilities, of smart locks 153
compromised device injecting malware 103
confidentiality, integrity, and availability (CIA) 102–103

consequence levels, of attack 15, **15,** 16
convolutional neural network (CNN) 29, 34, 38, 191
classifier 45, 46
hyperparameters **45**
CPS *see* cyber-physical systems (CPS)
credit card fraud identification techniques 196, *197*
critical infrastructure 6
cyber human systems (CHS) 199
cyber-physical systems (CPS)
autoencoders 193, *193, 194*
breakthroughs in 199–200, *200,* 201
critical in modern world 201–202
5C structure *200*
evolution of 202–204
in IOT devices 194–195
restricted Boltzmann machines model 191–193, *192*
structure 200
testability 201
cybersecurity 167

data anonymizing 215–217
data concentrators (DCs) 2, 10
data protection 208
DDoS *see* distributed denial of service (DDoS)
deep convolutional GAN (DCGAN) 33–34
adversarial samples 37–38
behavior feature extraction 40–41
convolutional neural network 38
data preparation 36
dataset 35
dynamic analysis 35–36
evaluation 47–48
experimental setup 39–40
generator and discriminator networks 45
hyperparameters **43**
image classification 43–44, *44,* **45**
image generation 37, *37*
synthetic images 41–43, **43**
words to images 41
deep learning discriminative adversarial networks 31
deep learning malware detection 28–29
deep learning methodologies 191, 197–198, 21
deep neural networks (DNNs) 31, 191
deep Q-network (DQN) 221
deep reinforcement learning 31
denial of service (DoS) 77
device attributes 91
device external process state 92
device internal process state 92
device provisioning gateway (DPG) 95
device provisioning service (DPS) 95
device state properties 91, 93, 94
digital twin 96–98, 102
distributed denial of service (DDoS) 11
domain generation algorithm (DGA) detection 195
DPG *see* device provisioning gateway (DPG)
DPS *see* device provisioning service (DPS)
DroidBox 40, 41
dynamical model 58
dynamic link libraries (DLLs) 27

e-commerce 191
edge computing 219–220
electricity service provider (ESP) 4
elliptic curve cryptography (ECC) 77
encryption 18, 24, 55
end-to-end encryption *see* hop-by-hop encryption
environment parameter 93
ESP *see* electricity service provider (ESP)
eventual consistency model 153–154
exchanging electronic keys 149

feature selection techniques 28
feed-forward neural networks 32
FHE *see* fully homomorphic encryption (FHE)
Fisher score technique 28

5G healthcare devices 167
flow authorisation 102
FRAmework for Sensor Application Development (FRASAD) 125
fully homomorphic encryption (FHE) 18

game-theoretic model 167
  active & idle probabilities 180, *182*
  cognitive communication systems 167
  competitive strategy 176
  equal transmission power 177
  long-term performance 180, *183*
  multiple users extension 179–180
  noncompetitive strategy 176–177
  outage probability 170–171, 180, *181*
  payoff matrix calculation 180, *182*
  performance analysis 177–178
  performance metrics 180, *184*
  problem statement 168–170, *169, 170*
  spectrum-sharing cognitive systems 167, 168
  uncoordinated transmission strategy *173,* 173–175
  zero-determinant strategies *171,* 171–173
game-theoretic transmission strategy 61–64, **65,** 68, 73
generative adversarial network (GAN) 209; *see also* deep convolutional GAN (DCGAN)
  accuracy measures 225
  convolutional framework 25
  data analysis and interpretation 224
  data preparation 223, *223, 224,* **224**
  F-measure 225
  incorrect classification 225
  metamorphic malware detection 33
  privacy 225, 226, **226,** 227, **227,** *227,* 228
  in privacy data analytics 220–221
  privacy measures 225
  third-party malware detectors 33

HAPT2015 dataset 221, **222**
Hidden Markov Models (HMMs) 28
home area network (HAN) 6
hop-by-hop encryption 216
hyper-V technology 38

IIoT *see* industrial IoT (IIoT)
image generation 37, *37*
industrial IoT (IIoT) 54, 165
Information and Communication Technologies (ICT) community 121
information security 55, 57
  game-theoretic transmission strategy 61–64, **65,** 68, 73
  generalized transmission strategy 66
  model dynamics 66–68, *67*
  simulated use 68, *69–72*
  system model *58,* 58–59
  zero-determinant strategies 59–61, 65–66
information security risk assessment (ISRA)
  definition 6
  identification 7
  model-based approach 9
  profiling 7
  qualitative risk analysis 8–9
  quantitative risk analysis 8
  risk management process 7
  stages *7*
  treatment 7–8
innovative technologies 54, 165
International Data Corporation (IDC) 121
Internet of Things (IoT)
  applications 1, *2*
  architecture and applications 209–210
  definition 54, 124
  limitations and challenges 210–212
  privacy 212
intrusion detection system (IDS) 18
inverse document frequency (IDF) 36
ISRA *see* information security risk assessment (ISRA)

keeping access logs 150
K-nearest neighbor (KNN) 196

Knowledge Discovery and Dissemination (KDD) dataset 194

legitimate user (LU) 58, *58,* 59
LEONORE 79
likelihood level, of attack 15, **15,** 16
Linux kernel system 29
locking/unlocking the door 149
Low Orbit Ion Cannon (LOIC) 104

machine learning (ML) 26–28
malcode/malware 24
MalGAN 32, 33
Man-in-The-Middle (MiTM) attack 103, 109, 152
Markov chain 60
mashup-as-a-service (MaaS) 124
MDE4IoT 125
MDMs *see* meter data management systems (MDMs)
medical CPS 201
metamorphic GAN system *36*
meter data management systems (MDMs) 2, 10
model-driven architecture (MDA) 122, 124
multi-factor authentication (MFA) 160
multipurpose binding and provisioning platform (MBP) 79

neighbor area network (NAN) 6
network infrastructure 77–80, 87, 89, 98, 114
Node-generator 137
Node-Red 137, *139*
Node-red-contrib-web-of-things 137
node-to-node network 18

Object Management Group (OMG) 128
outlier detection 196

PEE *see* provision evaluation engine (PEE)
physical-layer security 57
poisoning attack 30
policy-based secure device provisioning 90–91
policy-based security application (PbSA) 98–100
policy database (PD) 95
policy-driven security architecture; *see also* secure smart device provisioning and monitoring service (SDPM)
  challenges *89*
  device provision 89–94, *91, 93,* 110–111
  digital twin 96–98
  security provisioning protocol 95–96, *96*
Policy Enforcement Fog Module 219–220
polymorphic malware code 25
polymorphism 24–25
possible level, of attack 15
precision agriculture 76, 200
Privacy Policy Agreement (PPA) 219
privacy-preserving data generation 209
privacy-sensitive, situation-aware description model 128–129, *129*; *see also* Web-of-Things (WoT) Thing Description (TD)
  implementation 132
probability density function (PDF) 59
proof-of-concept (PoC) implementation
  network setup 100, *101*
  performance evaluation 109–111, *112*–114
  security analysis 101–104, *104, 105,* 106–109, *107–108, 110*
provisioned devices database (PDD) 94
provision evaluation engine (PEE) 95
public key infrastructure (PKI) service 10

rare likelihood level, of attack 15
reactiveness factor, of legitimate user 63, 66, *67*
receiver unit (RU) 58, *58,* 59
recurrent neural networks (RNN) 31
reinforcement learning attacks 31
relay attacks, of smart locks 152, *152*

Remote Authentication Dial-In User Service (RADIUS) protocol 80
restricted Boltzmann machines (RBMs) model 188–189
  cyber-physical systems using 191–193, *192*
  fraud and anomaly detection 196–199, *197*, **198**
  malware attack detection 195–196
  targeted problems using 189–191
risk management techniques **17**
risk profiling phase 7, 14–16, **15**, *16*, **17**, 211
risk treatment phase 16, 18

SDPM *see* secure smart device provisioning and monitoring service (SDPM)
SecSmartLock (Secure Smart Lock) 155
secure smart device provisioning and monitoring service (SDPM) *94*, 94–95, 95–96, *96*
  average path setup time 111, *114*
  device power consumption 111, *113*
  digital twin 96–98
  performance evaluation 109–111, *112–114*
  policy-based approach 90–91
  policy language for 91
  post-condition *93*, 93–4
  pre-condition *91*, 91–92
  requirements 90
  resource utilisation 111, *113*
  security policy specifications 99–100
  security properties 101–103
  throughput 111, *112*
security pillars 2
security policy specifications 99–100
self-driving cars 191
self-taught learning (STL) 191
service-oriented architecture (SOA) 125
shallow machine learning techniques 26
signal-to-interference plus noise ratio (SINR) 59
signal-to-noise ratio (SNR) 59
single loss expectancy (SLE) 8
single malicious packet 103
SituationPrivacy 129–130, *130*
SituationPrivacyWoTTD metamodel 131, *131*, 132, *132*
smart cities security 208, 210
smart homes privacy and security 155–156; *see also* smart locks
  authentication 160
  configuring device options 160
  data at risk in cloud 156
  data collection and mining 156–157
  device selection 159
  insecure devices 158
  mitigation strategies 159, **159**
  multi-user challenges 157–158
  network attacks 158
  network configuration 160
  physical safety 158
  self-censoring 159
smart locks 147; *see also* smart homes privacy and security
  access control 150
  attribute-based access control framework 154
  authentication and authorization 150
  components 147–148
  Device-Gateway-Cloud 148, *149*
  Direct Internet Connection 148–149, *149*
  eventual consistency model 153–154
  exchanging electronic keys 149
  keeping access logs 150
  locking/unlocking the door 149
  privacy and security 151–153, 155

privacy and security (*cont.*)
vacation mode 150
smart medication dispensers (SMD) 127–128, **128**
evaluation 138, *139,* **139,** 140
smart meters (SMs) 2, 10
software-defined network (SDN) 191
software defined provisioning (SDP) 80
spectrum-sharing cognitive communications 57, 167, 168
spoofing/masquerading 103
stacked autoencoders (SAE) 28–29
state consistency attacks, of smart locks 151–152
statistical disclosure control methods (SDC) 216
steered micro aggregaion 216
supervised learning 191

tenant privacy, of smart locks 153
term frequency and inverse document frequency (TFIDF) 36
Thingweb Node-WoT 137, *137*
two-way communication pathways 4

unauthorized unlocking, of smart locks 152, 153
unified modeling language (UML) 9
unmonitored biometric system 196
user attributes 91
user identification 211
user privacy 218
user tracking 211
utility monitoring and controlling 211–212

vacation mode, smart locks 150
variational autoencoder (VAE) 213
virtualization technology 38
virtual machines (VMs) 38
virtual private networks (VPNs) 160

wearable devices 208
WAN *see* wide area network (WAN)
Web-of-Things (WoT) Thing Description (TD) 122–123, *126,* 126–127, *127*
wide area network (WAN) 6
wireless sensor network (WSN) 166, 217–219
WoTT2SPWoTTD model transformation 133, *134–136*

zero-determinant strategies 58, 167
game-theoretic model *171,* 171–173
information security 59–61, 65–66

Printed in the United States
by Baker & Taylor Publisher Services